Making Sense, Making Science

Series Editor
François Rastier

Making Sense, Making Science

Edited by

Astrid Guillaume
Lia Kurts-Wöste

iSTE
WILEY

First published 2020 in Great Britain and the United States by ISTE Ltd and John Wiley & Sons, Inc.

ISTE Ltd
27-37 St George's Road
London SW19 4EU
UK

www.iste.co.uk

John Wiley & Sons, Inc.
111 River Street
Hoboken, NJ 07030
USA

www.wiley.com

Library of Congress Control Number: 2020941617

British Library Cataloguing-in-Publication Data
A CIP record for this book is available from the British Library
ISBN 978-1-78630-579-4

Contents

Preface

Progress - The image of Newtonian physics has enslaved
the divine sciences, then the humanities.
Why value the hard science?
They are so insecure, so fearful, so rightly modest.

François Vaucluse, *Épistémologie minimale*, 2019 (authors' translation)

This book, *Making Sense, Making Science,* is the summary of three colloquia held at Sorbonne University in 2015, 2016 and 2017 : "*Humanités et sciences de la culture*" (Humanities and Cultural Sciences)[1], "*Saussure et l'avenir des sciences de la culture, Vingt ans après* De l'essence double du language" (Saussure and the Future of Cultural Sciences – Twenty Years After *De l'essence double du langage*)[2] and "*Faire sens, faire science*" (Making Sense, Making Science)[3]. This last colloquium, conceived as an unofficial tribute to the work of François Rastier, was also intended to take stock of the progress made in the cultural sciences since the colloquium "*Textes, documents, œuvre (Autour de François Rastier)*" (Texts,

1 May 2015, symposium organized by Astrid Guillaume and François Rastier: https://www. fabula.org/actualites/humanites-et-sciences-de-la-culturequestions-d-avenir_66796.php (accessed July 23, 2019).

2 May 2016, symposium organized by Astrid Guillaume and François Rastier: https://www. fabula. org/actualites/saussure-et-l-avenir-des-sciences-de-la-culture-vingt-ans-apres-de-l-essence-double-du-langage_73571.php (accessed July 23, 2019).

3 November 2017, symposium organized by Astrid Guillaume, Lia Kurts, Franck Neveu and François Rastier: https://www.fabula.org/actualites/faire-sens-faire-science-colloque_81424. php (accessed July 23, 2019).

documents, works (concerning François Rastier),[4] which was held in 2012 in Cerisy-la-Salle.

Faire sens. De la cognition à la culture is the title of a work by François Rastier in which he presents his interpretative praxeology and cleverly distinguishes it from cognitivist approaches (Rastier 2018). It is interesting to note that he recalls Saussure's essential contribution to "pragmatic" thinking in linguistics, because this interpretation of the Saussurian corpus is still insufficiently disseminated.

Indeed, while the pragmatic or praxeological approach is spontaneously associated with Peircean theory, the Saussurian theoretical gesture, which consisted of "repatriating the signified in languages" and, jointly, taking as its central theme the "life" of signs within societies, naturally leads to the recognition that the signified and the signs themselves are the result of forgotten interpretative actions, depending on the context of interpretation, itself conditioned by the consideration of multiple parameters, notably the corpus of emergence. At the same time, this perspective thus leads us to recognize that the traditional distinction between semantics and pragmatics is irrelevant: linguistic signs are more or less decontextualized "passages" of texts. Moreover, it also means recognizing scientific meta-language as a responsible fact that is constantly concerned with finding the most relevant criteria for legitimizing interpretation.

This work aims to recognize that "in the scientific field the interpretative act is not only in the *terminus ad quem*, but also in the *terminus a quo*. It is in the sequencing of 'data', in the construction of facts, in the theoretical framework, in the selection of relevance, and generally in all the analysis *a priori*. Knowing how to take into account the false absence of the researcher in the analysis supposes never abandoning reflexive thinking. Consequently, it also makes it possible to maintain the hermeneutic attention necessary for scientific activity", as Franck Neveu reminds us in his contribution.

In addition, *Faire sens, faire science* (*Making Sense, Making Science*), from *cela fait sens* in French or "that makes sense" in English, implies various significant elements that are explained here[5,6].

4 July 2012, symposium organized by Driss Ablali, Sémir Badir and Dominique Ducard: http://www. ccic-cerisy.asso.fr/rastier12.html (accessed 23 July 2019); proceedings published in Ablali *et al.* (2014).

5 Lia Kurts-Wöste intends to publish a work that explores two new complementary semiotic notions, *significativity* and "making sense", with a transdisciplinary value, contributing to the current renewal of the epistemology of cultural sciences, notably by integrating an intersemiotic perspective (music): the analyses produced here are taken from this work currently being prepared.

6 Editors' note: the text below describes the problems encountered when proposing the title in English, from our native language of French.

1) First of all, the understanding and interpretation of this expression in French seems to depend on the homonymy and the great polysemy of the term "*sens*", from the Latin *sensus*, which signifies "direction", the idea of heading in the right direction "understanding, reason, intelligence", or Lorraine Franconian *sinnu* > *der Sinn* in German, the meaning, which will lead to *forcené* (the French word meaning "enthusiast"), the person who goes beyond (good) meaning. The word was first written in Old French as *sen* and was influenced in terms of spelling by the Latin term, due to its semantic proximity to certain uses. "*Cela fait sens*" thus re-explains the directional etymology of the homonym *sen* and breaks free from its strict semantic definition: "*cela fait sens*" does not necessarily mean it "has a meaning", in the linguistic sense of the term, but "what is interpretable".

2) From here, the positive axiological orientation of the expression "*cela fait sens*" or "*cela a du sens*", literally "that has sense", designates a reality testifying to choices which seem, depending on the case, "sensible", in other words, reasonable, valid, legitimate, interesting, relevant, timely, even significant – so not only with an element of content but also of scope. Its use meets the definition of *significativity* and interprets the guidance of a self-reflective or external evaluation.

3) The agentive value of the verb *faire*, "to do" or "to make" in English, is actually bipolar, since in the use of this expression, a reality can be recognized, one that can be:

- the result of a choice, for example, an interpretative action leading to a decision of which this evaluated reality is the effectuation;

- an object of evaluation, i.e. an interpretive action and its link with a normative dimension.

The Saussurian approach, based on the principle of a *dual essence* of language facts is a systematic that has taken note of the fact that "the field of knowledge (and that of the theory of knowledge) is [now] determined by the conceptual field of doing (and that of the theory of doing)" (Bouquet 1997, p. 33, author's translation). It is therefore a praxeology that has become unavoidable in the interpretation of signs within societies.

The conceptual field of pragmatics is therefore part of the scientific order, even if it is not a given for everyone: linguistics has sometimes fallen short of the critical but also practical expectations that one could legitimately expect from it from this point of view.

For Franck Neveu, it is about "the development of a reflexive point of view within the methodologies, which marks a break with a certain form of candid objectivism that has long prevailed, an objectivism according to which the sciences would be determined solely by their internal logic, with no regard for the formal and material structure of their discourse. Paradoxically, language sciences have largely developed this perspective, whereas one could expect this field to have a better understanding of the effect of discourse on scientific ideas".

Moreover, certain preconceptions continue to present themselves as unassailable evidence or as inescapable academic realities. For Franck Neveu, "it is indeed a question of rethinking the question of the academic territorialization of knowledge, which in many respects seems to be non-congruent with the cultural and scientific spaces actually occupied, saturated or only crossed by this knowledge".

The crucial point that unites all the contributions of this work is therefore the question of the place that a philosophy of science should occupy in a philosophy of culture, and, in particular, a philosophy of linguistics: "much has been philosophized about Saussure, to the point of making him a kind of philosopher of language, without realizing that the philosophy of language becomes a philosophy of languages, as objectified by linguistics, and thus a *philosophy of linguistics*", recalls François Rastier[7] here. With the question of pragmatics, it is, first of all, a practical philosophy and a deontology that we refer to, and not to the search for ontologies with universal and an-historical values. While recognizing the historical relativity of scientific knowledge, it is a question of not giving up the goal of rational and critical objectification, that is to say, a plural and responsible reasoning, which can, under certain conditions, be open to emotion, empathy and art, always following an ethical and deontological approach. Indeed, scientific metalanguage must be subjected to an evaluation that integrates different normative strata aimed at protecting it from certain excesses:

– either excess empiricity or lack of generality. For Franck Neveu, "the excess of empiricity and the lack of generality necessary for theory can be thought of as a denial of scientificity";

– or the absence of questioning about the "necessity" of science, where the challenge is to admit, according to a broader kanticism, that the "what can we do?" can be regulated by a "what can we hope for?". Such a teleological dimension in no way hinders but guarantees responsible objectification, especially when it comes to preserving diversity and specificities[8].

In the academic world, there are still obstacles and hurdles that hamper research and its results by excessive mono-disciplinarity or by closed-mindedness in the face of possible contact and hybridity between the sciences and humanities. The names have a lot to do with this: by opposing *sciences* and *humanities*, "hard sciences" and

---

7 All quotes by Rastier are translations.

8 It should be noted that this theoretical combination of teleology and concern for the specificities of individuals has also been worked on in the English-speaking field, in particular, by Bertalanffy (1968).

"soft sciences", the *humanities* have been labeled "non-scientific" and have lost credibility. In the past, the term Human and Social Sciences (HSS) was introduced to circumvent this discrimination and compensate for this lack of recognition. Indeed, the humanities and cultural sciences share with the other sciences their vocation of knowledge. They develop their own methods to objectify their fields of research, to describe the singularities of cultural objects, which are as multiple as they are changing in a world that has become internationalized in a short period of time. The cultural sciences are, however, sciences of specificities, contrary to the idea inherited from Aristotle that there would be science only from the general, even if it must be recognized, at the same time, that they are linked to traditions and norms that are inscribed in time and space. The authors of this book show the original ways in which it is possible today to renew their epistemology, drawing on authors such as Leroi-Gourhan, Merleau-Ponty, Simondon and others, while taking into account the changes in thinking and research in this new century (on statistics and cognitive approaches, for example). Cultural objects call for a specific program of knowledge, a rational study of the arts to develop new fields to be described with access to ever-expanding collections, whether digitized or not.

This book therefore also presents a reflective, critical approach – which can be called "semiotic" even when it does not explicitly use the term – to different types of cultural mediations (language and non-language arts, life sciences and technologies, ethics, esthetics, etc.). It is intended to take account of this complexity of cultural content in order to master fundamental dualities, such as content and expression, synchrony and diachrony, but it also integrates the verbal and non-verbal, the sound and the non-sound of the human and the non-human, whether animal or machine. Indeed, the 21st Century has witnessed the emergence in all areas of society, the consideration of the non-human (living: animals, plants, planet), the non-living (machine, humanoid or animaloid robot, artificial intelligence and its implications), and even hybrid forms (trans-animalism). The once omnipotent and ubiquitous human is increasingly being challenged by animals and robots that now surpass the "human", whereas computer entities already surpass the human cognitively: all ontologies are shattering. However, the question of a semiotic anthropology, which unites all the contributions in a more or less explicit way, attempts, at the same time, to underline the human's responsibility in an environment that surpasses them, conditions them and on which they vitally depend. It also highlights a type of complexity in invention and innovation (scientific and artistic) that cannot be reduced to a program: individuation, which combines, in order to be constituted, various parameters inherited through cultural transmission, and whose complexity remains inaccessible to the machine, particularly in taking into account the parameters of ethical regulation.

Content

This book is divided into an introduction, three parts and a conclusion. The chapters mentioned here testify, each in their own way, each with their own direction, to the fact that the cultural sciences have a bright future ahead of them. They also show how the extent to which the labeling of the fields of knowledge is sometimes artificial and academic, and how these labels can be as many obstacles to innovation and the development of knowledge itself. From this point of view, each author underlines what their theoretical or more broadly epistemological options are. Each contribution identifies, first of all, the pitfalls, an identification from which each one establishes a differential consistency of their own approach and their own conception of science, which presupposes taking into account a history of ideas and the adoption of an approach based on a comparative point of view, a historicizing approach presupposing a certain humility that is not at all contradictory to a high epistemological ambition.

In his introduction, François Rastier (CNRS-INaLCO) clarifies the notions of "individuation" and "interpretation" of cultural objects by recalling what is at stake in the recognition of cultural sciences today. He contributed to the debate by recalling Cassirer's fundamental theoretical gesture that transforms "spiritual sciences" into "cultural sciences" and allows the emergence of the thought of semiotic mediation. In the face of the various identitarianist aberrations that can be observed today in the field of culture and in the face of the hyperspecialization of the human sciences, which jeopardizes the very possibility of federative and globalizing thoughts capable of reflecting on their responsibility, Rastier invites us to reaffirm the choice of the innovative project of the cultural sciences further taking into account the duality between local cultures and world culture, between the individual and the particular, and between the general and the universal: the semiotics of cultures and cultural anthropology inherited from Humboldt and Cassirer become of major importance here. They allow us to highlight the fact that the humanities and the cultural sciences, in particular, can claim to generalize without erecting as universals the regularities they objectify, so as to take into account singular events.

The author recalls that the Saussurian theory of views combined into dualities has made it possible to modify the very notion of objectification and stresses the importance of the self-reflection of science through a philosophy of action or praxeology – thus restoring the dignity of the historical and comparative projects of Cassirer and Saussure against the unfounded attacks of Heidegger and Derrida. These projects were based on the principle of a possible union between scientific culture and "literary" culture, in the name of an emancipatory definition. Since the social dimension predominates in the cultural sciences, semiotics considers that signs can only be conceived in relation to the societies that institute them. Thus, beyond academic issues, the critical attitude of the sciences and the arts meet to overcome stereotypes of belonging and promote freedom of thought. If the cultural sciences have a political impact, it is through developing a project of knowledge involving human diversity inherited from the Enlightenment period.

Even though the cultural sciences respond to a federative project, they reject the attempt to unify the sciences as formulated by Morris and Carnap, insofar as this project leads to not recognizing the contribution of the cultural sciences' methodology by hypostatizing scientific criteria (causality, statement of laws) which are, moreover, called into question in the "hard" sciences themselves. The development of rigorous theoretical principles for studying societal phenomena and diverse and varied cultural objects is well under way. Different types of *sciences* have emerged within the HSS, including historical and religious sciences, which does not imply that they are exact disciplines in the mathematical or computerized sense of the term, but it does imply that in order to approach these sciences, it is necessary to develop reading grids and theories that allow scientific rigor to be applied to moving subjects, which are sometimes ideological, because they are linked to plural and evolving disciplinary contexts. Thinking about the cultural sciences against this dynamic backdrop becomes a challenge that requires the precise study of verbal and non-verbal, sonorous and non-sonorous languages, of the implicit structures of societal phenomena, of which we often perceive only a tiny explicit part. Language sciences could therefore have a key role to play, requiring prior recognition of their status as cultural sciences: science of conditions, not laws, of values, not facts, while avoiding substituting, as Franck Neveu reminds us, "empirical naivety has taken the place of theoretical naivety".

François Rastier finally calls for a morphological reconception of the concept of structure, faithful to the origins of the elaboration of the first structuralism. The descriptive richness of a morphologically inspired structuralism lies, paradoxically, in the economy of its regulatory hypotheses, which allows it to adapt to the metamorphosis of forms. The sign is conceived as the metastable result of a process of individuation. The question of morphogenesis testifies to its fertility by allowing us to introduce, into linguistics, the notion of continuum and dynamicist modelizations that are interesting (for speech) in the fluency of phonation, the variety of enunciative gestures, prosody, isotopies and isophonies, vocal and emotional tonalities, and, more generally, in the singular points favored as places of relevance. A theory of the operations of appearances of forms and their modifications could be called an *operatic*. Art occupies an eminent place in this framework, as the works demonstrate the creation of signs in an exemplary manner. Such an *operatic* is particularly interested in extended corpora and does not hesitate to use computer processing, especially statistical processing.

As for the status of semiotics, François Rastier reminds us that "semiotics has long been hesitating between two vocations: sometimes it defines itself as a discipline among others, sometimes as a kind of meta-discipline whose mission is to redefine within itself the whole of the cultural sciences, and even a good part of the natural sciences. Between the two, the image of a "pilot science" has gained support, because of the very ubiquity of the signs themselves. François Rastier thus participates in the debate on the epistemological status of semiotics in its relationship with the cultural sciences. He concludes that semiotics "cannot, of course, make them obsolete or replace them, since each contains semiotics that are more or less explicit, often 'at the practical level'", and that "it is up to semiotics to reflect on these particular semiotics in order to be able to unite them and strengthen their unity". In a demanding project of pluri-disciplinarity and interdisciplinarity, the relations of auxiliarization between disciplines are reversible and do not function in any way on the basis of relations of domination. If semiotics can function as a meta-discipline, it is only as a vector for critical self-reflection on the very status of the theories, with a strong institutional dimension, in each of the disciplines, which have particular semiotics.

The introduction of this book is followed by the first part, which deals with the "Semiotic Foundations of the Cultural Sciences", in which Jean Lassègue (CNRS), Arild Utaker (University of Bergen, Norway) and Vincent Bontems (CEA) present a renewal of scientific thought based on a fundamental questioning of the human being. Starting from a reflection on symbolic forms in Cassirer's work, moving to the paleoanthropological point of view of Leroi-Gourhan, and lastly to a reconsideration of the importance of technical thought in the years to come, based on the Simondonian approach.

To begin with, Jean Lassègue (CNRS) addresses the "symbolic forms" in Cassirer's thought. He places the conception of the organization of cognitive subjectivity in Kant's supposedly universal human knowledge at the origin of this concept and does so in order to underline the epistemological revolution constituted by the recognition of the consistency of the semiotics of cultures, through which the phenomena of transmission, transformation, translation and the pluralization of rationalities are conceivable. Lassègue, however, returns the authorship of the problem of the variety of rationalities in the tradition of transcendental philosophy (as shown by the need for the three *Critics* and the tripartition between transcendental, empirical and reflexive) to Kant, which is traditionally attributed to Cassirer. He shows that the latter is then indebted to the teachings of the neo-Kantian philosophers of the Marburg school who sought to re-found the transcendental project in order to be able to take into consideration the advancement of science during the 19th Century, essentially in mathematics and physics.

What Cassirer "invented", and what our thinking of scientific rationality is still indebted to him for, is the idea of a *potential transformation of the principles of knowledge*, as well as their relative permeability. Thus, each mode of objectification institutes a particular way of producing meaning, but the recognition of each of these modes of objectification is more a matter of accentuation than of homogeneity of the space of knowledge under consideration. The notion of meaning, and the question of the specific modes of production of meaning in each scientific approach mixes the registers of the pure and the empirical – which would seem to be strictly contradictory in Kant's eyes. Lassègue then presents the concept of the "transformation group" or the statistical method as enlightening examples of the progressive diffusion that is possible across all fields of knowledge in a non-unidirectional manner (from mathematics to linguistics or psychology for the former, from the human sciences to gas physics, and back to sociology for the latter). Faithful to this transformational principle, Lassègue finally proposes a new definition of the vocation of science based on a consideration of the history of statistics, a definition that Cassirer was not in a position to think about but that he would probably not have denied: science aims to make explicit *the local conditions of possibility* of a certain regime of knowledge.

Then Arild Utaker (University of Bergen) develops the problem of meaning in paleoanthropology in Leroi-Gourhan's work, in particular, with the construction of a collective memory. Utaker underlines the fact that, without critical distance, human paleontology (paleoanthropology) risks being nothing more than a projection of our current conceptions of humanity. As a philosopher of language, he considers that the revolutionary aspect of Leroi-Gourhan's approach lies in the fact that it dispenses us from starting with thought and logic to define the human. With regard to "thought", Utaker recalls that Leroi-Gourhan had the merit of considering that it is up to "fossil men" to overturn the dominant image of humanity. What criterion should define the human race? Leroi-Gourhan is formal. The criterion is given by the morphology of the body: humans are defined by standing, and it must be admitted that the human species "started with our feet".

Without repeating all of Leroi-Gourhan's demonstration, Arild Utaker starts from the idea that the mind and language – whose development is the consequence of the development of the brain, because of the liberation of the hand and the elaboration of increasingly complex technical gestures that it has enabled – are conceived as responding to the same "function", which is primarily a function of transmission and preservation of the collective memory. Humans thus emancipate themselves from genetic memory and become cultural beings. Utaker defends, from the interpretation of Leroi-Gourhan's work, a complex position, typical of the cultural sciences, which seeks to avoid both the complete biologization of cerebral capacities, characteristic of orthodox cognitivism (which goes hand in hand with a mechanistic conception of life proper to the neo-Darwinian paradigm) and the equivalent forgetting of the body, in particular, the hand and the technical gesture, where technique becomes subordinate to thought and where our philosophy ends up oscillating between mentalism and cerebralism. If language and the brain respond primitively to a function of transmission and memorization of technical gestures, the brain is not an organ like the others, it is the organ that connects and *coordinates* functions: locomotion, nutrition (and grasping) and relation (perception) are the three major functions. Following the example of Leroi-Gourhan, who proposes to speak of "syntax" to designate this capacity for coordination and articulation of functions, Arild Utaker proposes to consider that Man is first and foremost a "grammatical man", considering that the "neuronal man" of neo-Darwinians and traditional cognitivism cannot explain this complex coupling function, where tradition is taken into account, insofar as the anatomical body, for its part, follows the mechanics of cause and effect.

As far as logic is concerned, Utaker underlines the importance of prelogical thought in Leroi-Gourhan's paleoanthropological approach and recalls the link he establishes between multidimensional graphics and mythology, refusing to consider that pictography is a form of childhood writing.

Finally, Vincent Bontems (CEA) addresses the difficult question of language in its relationship with technical thinking. Based on the Simondonian conception of technology and its relationship with language, he specifies some original points of convergence that can be thought of today between the science of technology and linguistics. First of all, he notes that Simondon's work does not contain any in-depth analysis of language and he recalls the reasons for this absence: for Simondon, language is not the original condition of meaning. On the contrary, Simondon insists on the need to take into account multiple channels and technical functions of communication, of which articulated language is only one particular mode and not the paradigm. More than any speech or text, it is the technical object that is the material and symbol of transindividuality because it universally conveys information.

Bontems acknowledges that the radicality with which Simondon dismissed the linguistic paradigm and reduced the anthropological scope of language inquiry may well offend the sensibilities of some linguists. However, he emphasizes how much the orientations and demands of his research fundamentally converge with those of such a rigorous linguist as François Rastier, who was committed to dissolving, in a Saussurian spirit, the ontological preconceptions of language and to denouncing the Heideggerian mysticism of Being.

Far from denigrating the work of the linguist, the Simondonian philosopher who assumes the demands of rigorous technology must therefore seek help in articulating his own investigation with the terminology of engineers. This is the meaning of the collaboration between V. Bontems and linguists on Safran's inertial unit lineages.

In return, research to operationalize the "diagrammatization" of technical patterns and lineages undoubtedly suggests a new field for the "science of signs within societies". Bontems finally gives the example of a possible application of

Simondon's approach in "genetic mechanology", in collaboration with engineers. Once the position of the "Micromegas" lineage within the family of gas ionization detectors had been defined, it shows that he was able to trace his genealogy by identifying the inventions that gave rise to technical lineages realizing, at least partially, the same technical scheme.

In the second part of this book, "Hermeneutics of Science, Hermeneutical Sciences", Franck Neveu (Sorbonne University), Régis Missire (University of Toulouse) and Astrid Guillaume (Sorbonne University) reflect on the importance of mathematics, implicit forms or geometry to renew thought patterns in a multidisciplinary or intertheoretical perspective generating new content. In order to analyze this innovative and changing content with a semio-anthropological compass, allowing themselves to adopt approaches that are not only anthropocentric, they clarify the status of theories and descriptions so that a true interdisciplinarity around new observables, both within the cultural sciences and in cooperation with other scientific departments, can emerge. It is the status of interpretation and the importance of definitions that are also at stake in this chapter: which hermeneutics and which terminologies for which sciences of tomorrow?

The aim of Franck Neveu's study is to examine the link between philosophy and mathematics, and, more broadly, between the spiritual sciences and the presumed exact sciences. More precisely, he wonders how a philosophy of mathematics allows us to think of original links between linguistics and mathematics.

He works on the field of the philosophy of mathematics and on the common origin of the two fields (philosophy, mathematics), but also on the fact that mathematics is a precondition for the existence of rational philosophy, in particular because it allows us to question universally true judgments. The historical separation of mathematics and philosophy in the 19th Century foreshadowed the epistemological disruption between the "natural sciences" and the "human sciences": therefore proving fundamental. Cavaillès' work, which conceives of mathematics as a future, allows us to think differently about the relations between these two scientific fields. "On the reasoned emergence of these two approaches, we can only refer to Wilhelm Dilthey and his *Introduction to the Human Sciences* (1883: *Einleitung in die Geisteswissenschaften*), where he takes a stand against Comte's positivism, whose theories are distorted at least by the fact that there is not one method of knowledge, but two: that of the natural sciences, and that of the humanities, or sciences of the mind". While Franck Neveu does not specify this in his contribution, we point out in the introduction that it was not until Cassirer, however, that the "spiritual sciences" were reconceived as "cultural sciences". This terminological reworking actually involved a complete reconception of the sciences, including the natural sciences, as specific mediation practices. Whether in Cavaillès' or Cassirer's work, it is the problematic status of mathematics that raises the question of the need for such an extension and the recognition that each time, even if in different ways, it is a question of the complex coupling of the human being with their environment.

For Dilthey, what prevails in the natural sciences is the explanatory, therefore causalist and nomological, method; studying phenomena whose intelligibility is extrinsic because it depends only on the regular relations that these phenomena maintain with other phenomena whose effects they are. What prevails in the spiritual sciences is the comprehensive method, which is interested in the conditions of the emergence of phenomena and links them to their issues. The Diltheyan opposition has been overcome by Cassirer who redefines the spiritual sciences as cultural sciences.

Thus, in this history of the relationship between philosophy and mathematics, the crux of the problem is not, as one might expect, that of highlighting the logical formal, but, counterintuitively, that of taking historicity into account. Mathematics and philosophy thus separated in the 19th Century in the name of taking into account the theme of becoming, according to a movement that began with the French Revolution, where concern for history valued movements, revolutions and negativity, at the expense of the *subspecie aeternitatis* contemplation of mathematical truths, which were considered timeless as soon as they were established. From this separation, it is possible to identify two joint pitfalls, each in a different field: Neveu shows that it is a question, on the one hand, of overcoming the pitfall of a historical universalism, which Cavaillès' philosophy of mathematics shows how to avoid, by working on the mathematical future and on "understanding", considering that this theme is not restricted to "human sciences" but rather by implicitly recalling that all science is made by humans and therefore, as such, is "human science". On the other hand, overcoming the pitfall of a simple account of historical facts, to which Cavaillès reduces the human sciences, in a partial and questionable manner, not to mention irrationalism, exemplified, in particular, by the Heideggerian existential hermeneutics, which Neveu adds to the list. Cavaillès thus faced a double challenge: his aim was to escape both the apriorism of reason (as well as the apriorism of Being) *and* the empiricism of history (by assimilating, in a highly problematic manner, the historical approach to a simple statement of facts). In any case, he posed a question that is still unanswered: what means can we give ourselves so that the refusal of a naive positivism does not amount to renouncing all forms of positivity?

Franck Neveu concludes by showing that it is therefore possible to have the same epistemological approach to mathematics as to linguistics and to consider that both operate on categories of thought, so that a science of understanding mathematical categories of thought or linguistic categories of thought is possible: it can only be achieved by taking into account the history of disciplines and the history of ideas, according to a critical and comparative approach specific to the cultural sciences (even if Franck Neveu does not use this term). The spiritual sciences (Dilthey) include both mathematics and linguistics, according to Neveu, and are at least as much a science of the "mind" (a word to be understood here as a translation of *Geisteswissenschaften*) and its categories of thought as a critical distance from science conceived as "made by the mind".

Régis Missire (Jean Jaurès University, Toulouse) is interested in the semiotic articulation of textual meaning and what semantics based on a semiotic approach to language resources can say about it. Since Benveniste, who distinguished between *semiotic* and *semantic* levels, "the articulation of the semioticity of linguistic units with the thematico-intentional, praxeological and hermeneutic dimensions of meaning is problematic, to the point that the 'hiatus' between these two levels justifies for him envisaging the existence of two distinct linguistics, each conceptually matched in a specific way". It is from a critical interpretation of this approach that the continuity between meaning and linguistic semioticity has developed. At the same time, Missire reviews the proposals made in an attempt to carry out this analysis in its entirety within the semiotic framework (on the basis, among others, of the work of Coseriu, Hjelmslev, Lyotard and Rastier). He quotes Rastier (2003, p. 42) for whom "[...] the mysterious exterior of language is only the other side of the logico-grammatic restriction of linguistics: it resides in the texts and the intertext that it cannot conceive". By repatriating the "outside" of language into the intertext, and thus breaking away from the referentialist tradition of semantics, Missire proposes to overcome the theoretical dichotomy put forward by Benveniste and to consider the semantic regime as conditioned by the semiotic regime of texts. Missire is interested in two aspects of the semiotic articulation of textual meaning: first, he distinguishes three semiotic relations that constitute textual semiosis (*meaning/designation/expression*) by distinguishing the types of units they relate to at the two levels of language. He then discusses the importance of the notion of *significance* in interpretative processes. This approach allows him to propose a definition of *textual meaning* that consists of "(1) the correlates, in terms of the signified, of the relations of meaning, designation and expression and (2) the relations between these correlates". Similarly, he defines textual *semiosis* as "the set of relations of meaning, designation and expression, as well as, for each of them, the possibility of a relation of significance". He proposes that, at both levels of language, the correlates of the relation of expression can be described "with the concepts of semantic and expressive backgrounds and forms" (Rastier 2015), which makes it possible to reinvest at this level "the descriptive concepts of the relation of meaning (notably the concept of seme) to describe units (notably semic molecules) that are no longer linked to their correlate on the signifier plane by a reciprocal presupposition relation. Conversely, the expressed can be converted into the signified when it enters into a relation of signification associated with a semiolinguistic unit".

Thus, for Régis Missire, it is always necessary to distinguish in a text what an *extract* can be understood as *signifying*, *designating* or *expressing*, because the textual level includes relations of signification, designation and expression. Finally, he reminds us that "access to meaning is at best asymptotic, and that no discipline can exhaust its description".

Astrid Guillaume (Sorbonne University) works on interdisciplinary theoretical hybridizations, i.e. on the transfer of theories from one field of knowledge to one or many other fields in order to renew thinking and lines of research in HSS. She seeks to decompartmentalize academic disciplines. Referring to the definitions of the physicist Basarab Nicolescu, Astrid Guillaume recalls the meaning of the terms, frequently but often erroneously used, of *pluridisciplinarity*, *interdisciplinarity* and *transdisciplinarity* in order to define the notions of *pluritheoricity*, *intertheoricity* and *transtheoricity*.

The aim of her project is a theoretical innovation that opens up new disciplinary fields, accompanied by neologisms or updates of terminologies and speciality languages, so that language sciences can keep pace with the progress made in the life sciences and technologies over the last 20 years. Intertheoricity (or the hybridization of theories) here enables a meeting of fields of knowledge, opening up the HSS to the life sciences (zoosemiotics and biosemiotics), the technological sciences (transhumanism, transanimalism, humanoid and animaloid robotics and artificial intelligence) and other disciplines in the HSS (transferogenesis, semiotraductology and translationogenesis). By working on intertheoricity, Astrid Guillaume achieves a transtheoricity that allows programs of ethical universalism such as humanimalism, a humanism that goes beyond the humanist and animalist spheres, which are often ideological, because they are part of an ontology of the human and animal. In doing so, she also breaks up ontological approaches that have been established for centuries: "Intertheoricity opens up disciplines and blurs overly rigid disciplinary outlines; it breaks down the boundaries of the fields of human knowledge by bringing together the humanities and the sciences, but also the arts and spiritualities of the world, long sidelined by scientific and theoretical approaches".

Translation, at the heart of these hybrid transfers, whether theoretical, lexical, semantic or textual, then takes on its full meaning. The theoretical transfer can only take place at the end of a triadic process of composition/decomposition/recomposition, which is not unlike the verbal translatological transfer (verbalization/deverbalization/reverbalization): it responds to three characteristics: *elasticity*, *plasticity* and theoretical *hybridity*. This inter-theoretical semantic transfer paves the way for new theories that make it possible to analyze current social phenomena that are particularly plural, multifactorial and multidisciplinary.

Art and geometry, taken, respectively, as a model of hybridization and modeling itself, are a source of inspiration to be fully integrated in the processes of theorization and theoretical transfer. The artist Grégory Chatonsky speaks of "leafing through realities" involving a multiplicity of interpretations. Thus, Guillaume reminds us that what makes the theorization of human and non-human cultural activities complex is precisely this multiplicity of interpretations that can be made of them. In order to take into account this polysemy of the cultural fact, to maintain scientific rigor in the study of signs within societies and to carry out analyses that are as detailed as they are objective and complete of the fields of human and non-human knowledge, Astrid Guillaume calls for a return of diachrony within the academic curricula and all fields of specialization of knowledge, so that the cultural sciences make sense and science in synchrony as well as in diachrony. Theories, whatever the disciplines to which they relate, are often limited temporally, they are embedded in contexts and histories. At this time, for example, parts of Newton's theories are being challenged by black hole research. But they are not outdated, they are embedded in other theories that deal with phenomena that Newton did not evoke. Animal intelligences and thoughts are scientifically proven by ethologists and zoobiologists where many contemporary philosophers and linguists still deny them. This reopens the debate on interdisciplinary polysemic definitions and legitimately paves the way for zoosemiotics. Theories thus have their own historicity; this implies the perspective necessary to analyze them objectively and to recognize their limits when technological progress allows them to evolve, or even to exceed them. The history of theories and sciences takes on its full meaning here.

In this context of the renewal of knowledge and the questioning of theories of the past, sometimes still anchored in the curricula, it will be up to the academic structures evaluating the HSS to be more flexible in order to "be able to adapt to these new thoughts, and not the other way round".

In the third part, "Literature and Arts Sciences", Lia Kurts-Wöste (University of Bordeaux) and Pierluigi Basso-Fossali (University of Lyon II) are interested in the stakes of a non-logocentric semiotics of cultures based on explorations on music and *significativity* and on the roles of a semiotics of the arts that makes it possible to overcome the shortcomings of the past.

Lia Kurts-Wöste committed to developing the descriptive vocabulary of the semiotics of cultures from the off-center point of view of music. She legitimizes such an approach by pointing out that most of the authors that constitute the reference base of cultural sciences today have defended a non-logocentric approach (Leroi-Gourhan, Simondon, Cassirer or even Saussure – to name just a few major figures). This rejection of logocentrism takes different forms each time, which she declines precisely. She then notes that it was at the same time and in the same geographical area (19th Century Germany) that the characteristics of Riemann's musical hermeneutics and Schleiermacher's material hermeneutics were defined. After recalling the characteristics of the interpretative praxeology developed by Rastier's semiotics of cultures and its possible extensions for a general hermeneutization of the sciences, she proposes to explore the means of strengthening its non-logocentric character by relying on the heuristic value of musical hermeneutics and, in this way, to elaborate an original semiotic notion, that of *significativity*, which is absent from reference dictionaries in semiotics, which more readily archive the notion of "significance" from Benveniste. It should be noted that this notion also appears in adjectival form (the adjective "significant"), but in a more temporary manner, in the contribution written by Pierluigi Basso-Fossali, when he quotes Geertz and Goldwater. Lia Kurts-Wöste starts from Cassirer's position that it is never a question of reducing art to language or language to art, although, at the same time, it is recognized that art cannot be taken further without a linguistic approach and language cannot reciprocally be understood in all its depth without an artistic approach to culture.

Using the example of Adorno's musical analyses, she proposes to think of *significativity* as the majority regime of semioticity in music. *Significativity* thus designates *the critical interplay that a work constructs with the cultural dimension, and, in particular, with the inherited forms and norms.* The retroactive interest of such a non-logocentric definition for approaching the semiotic functioning of the verbal itself consists of the possibility of defining a triple regime of semioticity for the verbal (meaning/signification/*significativity*), and of adding the term "stake" to the list of terms proposed by François Rastier to describe the interpretative act (content, scope, address, destination, point of view, guarantee). Thus, the notion of *significativity* subjects the reconstruction of an intensity to a guarantee, within an epistemological framework that ultimately makes it possible to link a semiotics of singularities to a semiotics of norms and an ethics of plurality. The definition of such a regime of semioticity remains programmatic for the time being, and calls for further development and clarification, particularly in contrast to the existing typologies of the unsaid and the implicit, or to the linguistic notions of "significance" and "context". However, it already allows us to consider that a science of the arts conceived as a hermeneutic of *significativity* would thus constitute a fruitful solution to the "insoluble nature of this problem of the relationship between science and ethics" (Simondon), in the sense that it would achieve, in an *interpretative way*, a coherent and unified summary of theory and practice, by which the "encounter of technicality and respect with regard to totality" would not be reduced to a banal "ensemble of procedures and a mythology" (Simondon). Finally, a hermeneutic of *significativity*, centered on the arts, would underline the interest of defining the human being less by verbal language than by their capacity to handle, in verbal language and in other non-verbal semioses, *significativity* in emancipatory terms (Cassirer), whether in the arts or in the sciences.

Pierluigi Basso-Fossali develops what has been and what could be a semiotics of the arts, taking into account a wider cultural economy. After having reconstructed the different approaches to art in the esthetic and semiotic fields to underline a certain number of shortcomings or gaps, both in terms of reductionism (self-televism, immanentism, excessive valuation of the singular that does not take into account the work's place in a tradition, etc.) and "thaumaturgic" idealisms (referring to supposedly universal idealities (truth, beauty, harmony), the entire power of the art world). Pierluigi Basso-Fossali proposes to take part in the legacy of the cultural sciences, whose significant elements he recalls, while giving a current rereading in the light of the notion of "forms of life". The cultural object's life form, and, more specifically, of the work, requires a science of complexity, insofar as this life form is defined as a relational crossroads, creating tension, between three spaces of relevance (enunciation, implementation, instantiation), which are as many fields of integration (material, communicational, reconstructive). This complex modeling makes it possible to highlight some key elements of a new method of approaching the work in semiotics, an "ecological" approach, which allows us to escape the rut of analytical esthetics or neuroesthetics, often exploited by the semiotics of the arts. While "analysis" separates the spaces modeled here, "interpretation" seeks, on the other hand, to triangulate the fields of integration. Such a hermeneutic approach may give rise to reticence insofar as it does not guarantee homogeneous and repeatable results and requires an uncomfortable effort to problematize the relations between the instances concerned. However, it has the merit of being an invitation to awaken perceptive potentialities while considering the interpreter as a "responding" subject, far from a definition of semiotics as simple "sign and code sciences" that has sometimes allowed esthetics to discredit it.

This definition is based on the elaboration of a complex posture on the part of the interpreter, which responds to numerous requirements simultaneously, a difficulty proportional to the declared ambition of a science of the arts conceived as a responsible and sensitive interpretative act, which admits that art can question the meaning of science. Among these prerequisites, the following can be mentioned: recognition of the importance of the hermeneutic paradigm based on the achievements of the philological tradition; the need to inscribe works in corpora (diachronic studies are still little developed in semiotics); the interest of a comparative approach and the fruitfulness of the questions born of the ethnology of art; abandoning textualism and recognizing the specificity of the semiotic modes of functioning of the verbal and non-verbal arts (even if the possibility of transfers is emphasized at the same time); to claim the inactuality of the work that allows its critical position insofar as it proposes new conditions of thinkability, the appellative aspect of art being then recognized as a driving force of history; to admit the impossibility of a totalization of a landscape of meaning; to seek a difficult posture stretched between *Stimmung* and distance; to recognize the retroactive effect of art on the science of the arts and, more generally, on the humanities insofar as it constitutes a privileged observation post of the phenomenon of interpenetration between the observation system and the environment ("science in the *Humanitas* is only responsibility squared, *res gestae* that insist on other *res gestae*"). Cavell, Cassirer, Grosse, Boas, Goddman, Belting, Descola, Geertz, Goldwater, Jullien, Ingold, Rastier, Vico and Mélandri are some of the major references in the elaboration of this ambitious program that allows us to rethink the plural roles (critical and federative) of a semiotics of the arts.

Finally, in the conclusion, Bernard Reber (CNRS-Sciences Po) addresses the deliberation between science and ethics with regard to human communication capacities, as well as the way of considering, together, the dimensions that make science and those that make sense, understood as science and ethics, which implies reorganizations, epistemological choices and explanations to make the requirement of participation and rationality hold together. "In fact, many of the problems that arise, whether in science for political decision-making, in the case of health and the environment for example, public scientific controversies, for example energy choices or bioethical problems, or even what has been quickly classified as post-truth (Holzem 2019), require a decompartmentalization of the sciences of the being and the sciences of the ought to be. This opening up is accompanied by the formidable problem of a crossed, interdisciplinary epistemology, but also on the two levels of truth and goodness or, to put it another way, of science and ethics. If these two requirements no longer exist, then it is difficult to make science and even less so to make sense". Of course, one can only welcome the fact that State territories have abandoned their implicit preconceptions and have, in a way, become de-delegitimized, for the greater benefit of science, especially with the now constant calls for interdisciplinarity. However, the separation between scientific culture and "literary" culture persists, and for half a century it has had harmful effects on the HSS: either scientism with a candid positivism linked to the naivety of the existence of "data" or irrationalism.

Bernard Reber clarifies the resources of argumentation in the face of other communicative skills, such as narration and interpretation. Thus, "argumentation when it is complete includes different moments, which can be explained, evaluated, and which may or may not lead to agreements. Argumentation not conceived as a weapon for contradictory debates, but to be constructed, it would reduce asymmetry, where it is often the implicit and the reasons imposed as obvious, when they are false, that benefit the weakest audiences". To take into account the fact that there are different communication capacities is to take into account the fact that there are different ways of making sense and making science.

To conclude, our work intends, in an underlying or more assertive way, to continue, in a Casaresian spirit, the exploration of the different modes of objectification at work in the sciences, their pluralization at the same time as their common, interpretative and mediatizing status, and, *in fine*, their relation to ethics, where science is connected to societal, sociopolitical and environmental issues and to the question of scientific activity reconceived as a responsible act, whose critical, even eco-committed stance is the *sine qua non*.

This book questions disciplinary divisions, shows other possible configurations in history, and underlines their past and future exchanges, considering that disciplines sometimes separate academic boundaries more than scientific ones. In doing so, each contributor submits to a largely self-reflexive, historicizing and comparative posture for their own field, seeking in all cases to make *sense* in order to make *science*.

Astrid GUILLAUME
Lia KURTS-WÖSTE
July 2020

References

Ablali, D., Badir, S., Ducard, D. (2014). *Textes, documents, œuvre: perspectives sémiotiques*. Presses universitaires de Rennes, Rennes.

von Bertalanffy, L. (1968). *Théorie générale des systèmes*. Dunod, Paris.

Bouquet, S. (1997). *Introduction à la lecture de Saussure*. Payot, Paris.

Holzem, M. (ed.) (2019). *Vérités citoyennes. Les sciences contre la "post-vérité"*. Éditions du Croquant, Vulaines-sur-Seine.

Rastier, F. (2015). *Saussure au futur*. Les Belles Lettres, Paris.

Rastier, F. (2018). *Faire sens. De la cognition à la culture*. Classiques Garnier, Paris.

Vaucluse, F. (2019). Epistémologie minimale. *Plastir*, 53 [Online]. Available at: http://www.plasticites-sciences-arts.org/PLASTIR/Rastier%20P53.pdf.

Introduction

Semiotics of Cultures and Cultural Sciences

The field of culture is now divided by political currents that claim to be based on identity-based culturalism, such as American cultural studies, which focus on race or gender identities or on Russian culturology, which emphasizes ethnic identity. In these circumstances, the cultural sciences' innovative project must be reaffirmed and taken further to take account of the duality between local cultures and world culture: the semiotics of cultures and cultural anthropology are of major importance here.

Beyond academic issues, the critical attitude of the sciences and the arts come together to overcome stereotypes of belonging and promote freedom of thought.

Very pronounced in France for academic reasons, the separation between scientific culture and "literary" culture has harmful effects: it favors not only neo-positivist currents, but also anti-rationalist ones, such as deconstruction, which revile the sciences, dismissing the very notion of truth and placing the pleasure of the text above all else. Culture cannot be divided and cannot be based on exclusion.

The humanities and cultural sciences naturally have a vocation of knowledge that they share with the other sciences. They develop their own methods to objectify their fields of research and describe the singularities of cultural objects.

Their contemporary epistemology is in need of renewal, with authors such as Cassirer, Leroi-Gourhan, Merleau-Ponty and Simondon. Moreover, the very project of structuralism, in its methodological requirement, has not failed in any way.

Introduction written by François RASTIER.

Cultural objects obviously call for a specific program of knowledge, and a rational or even scientific study of the arts is now finding new fields to describe with access to ever more extensive collections, whether digitized or not.

The arts benefited, long before the sciences, from the illusions now described by the neurosciences. The interpretation itself obeys philological constraints and the description of meaning is guided by the letter.

This study takes the side of a semiotics which, in order to account for the complexity of cultural formations, knows how to master fundamental dualities, such as content and expression, synchrony and diachrony. This program, clarified by the discovery of new manuscripts by Saussure and what has been called neo-saussurism, is now asserting its scope – well beyond the considerations arising from the sciences of communication and cognition. Clarifying the status of the theories and descriptions will allow for true interdisciplinarity around new observables, both within the cultural sciences and with other scientific departments.

Semiotics has long been hesitating between two vocations: sometimes it defines itself as a discipline among others, and sometimes as a kind of metadiscipline whose mission is to redefine within itself the whole of the cultural sciences, or even a part of the natural sciences. Between the two, the image of a "pilot science" has gained support, because of the ubiquity of the signs themselves.

Here, we wish to contribute to the debate on the epistemological status of semiotics in its relationship with the cultural sciences: it cannot, of course, make them obsolete or replace them, since each contains semiotics that are more or less explicit, and often "at the practical level". It is up to semiotics to reflect particular semiotics in order to be able to unite them and strengthen their unity.

I.1. The renewal of Saussurism

I.1.1. *Saussure's boldness*

Saussure's methodological reflection favored an epistemological renewal. On the one hand, he took a critical starting point in the radical challenge of unquestioned beliefs and prejudices that abounded in the grammatical tradition. This presupposed a unified reflection rejecting inconsistencies as well as eclectic compromises. On the other hand, Saussure's methodology excluded the naive metaphysics of reference and all other considerations external to language, to be based on the description of the languages themselves, as seen, for example, in the masterful theory of the syllable. In other words, Saussure seemed to draw from his objects, languages, the very principles of their description: this was neither positivism nor inductive empiricism, however, because he did not proceed by simple generalization, but, through the theory of points of view combined into dualities, he modified the very notion of objectification – which implies a break with traditional ontology, through what has been called a *deontology*.

The scope of his theoretical gesture was immense, since it concerned the very status of theory – such as linguistics and cultural sciences have "borrowed" from the natural sciences, or even from the logical-formal sciences. Much has been philosophized about Saussure to the point of making him a kind of philosopher of language, without really realizing that the philosophy of language becomes a philosophy of languages, as objectified by linguistics, and thus a *philosophy of linguistics* (Rastier 2015, pp. 221–257).

However, his theoretical gesture went beyond languages to cover all sign systems, so that the very project of semiotics derived from it and therefore became necessary. This was not a definition of new fields (all sorts of particular semiotics have existed for a long time), but rather their consideration from a unified point of view. This is why semiotics can be considered both as a discipline and as an organon for the cultural sciences as a whole, without these two definitions contradicting each other.

By a benign retrospective illusion, one can also see in Saussure's project a programmatic epistemology, but Saussure was not a prophet and remained the bearer of a radical demand rather than a detailed "program" in due form. He neither anticipated nor called for the various currents that claimed to be his, even to the neo-Saussure of our days, and perhaps he would be severe toward them. Anyway, after a century, the fruitfulness of his dissident principles remains intact and deserves to be better exploited.

I.1.2. *Critique of models*

Are "semiotic models", first and foremost, in the Greimasian tradition the "semiotic square" and is the fundamental narrative structure universal? The neo-Grammarians – like today's generativists who succeeded them – saw "linguistic laws" in the image of physical laws. This is precisely what Saussure rejected.

Semiotic models are logical concretizations, or even simply graphical representations of such fundamental laws – in which we have seen cognitive principles, or even more general mathematical morphologies that would justify a "naturalization" of semiotics. In this hypothesis, it would be legitimate to encounter them, or even to project them everywhere, since laws of this nature owe their luster to the ubiquity of their verifications.

The question remains open, and I would gladly plead for a cautious attitude: the cultural sciences are historical and comparative; they can claim to be general without erecting the regularities they objectify as universal. Moreover, one cannot conclude from the general to the universal, especially since cultures are very unevenly documented and ethnocentrism has not disappeared.

I.1.3. *Semiological issues*

To define semiology, we refer first of all to the paragraph from the *Cours de linguistique générale*[1] devoted to it. However, if we consider the handwritten sources and the students' notes, we can see that the editors of the *Cours* have retained and in fact imposed a restricted conception of semiology, making it a science of sign systems, which remains compatible with the grammatical conception of language as a system. This conception has largely prevailed and the introductions to semiology willingly list these systems (games, uniforms, road signs, etc.).

Such an additive conception of semiology is very reminiscent of the *Tractatus de signis* and other treatises that followed one another over the centuries until Peirce's work. However, the ambition of historical and comparative linguistics goes beyond the description of grammatical systems, as it contributes to the project of a general anthropology whose lineaments Humboldt traced.

Yet, in reflecting on the relationship between the two extremes of the duality between the social and the individual, Saussure emphasized that all sign systems are institutions: language is one of them, unique in its kind, writing is another, and so on. Since the social dimension ultimately predominates, the different sign systems can only be understood in relation to the societies that institute them. It is therefore to the project of an anthropology – no longer philosophical, but historical and comparative – that semiology must be related. It derives in fact from the Humboldtian program that Saussure radicalized and is in line with the wishes of the author, who confided that only the almost ethnographic aspect of languages was of interest to him. He thus obliquely denounced the issues of the neo-grammarians who confined themselves to an "internal" study of languages and thus prefigured the chomskians of today.

---

1 The *Cours de linguistique générale*, published in 1916, 3 years after Saussure's death by two of his colleagues, Charles Bally and Albert Sechehaye, was developed using student notes. It has long been an authority, but after a study on Saussure's manuscripts, at present it is considered rather unreliable, even largely questionable (see Rastier 2015).

I.2. The need for structuralism

I.2.1. *Structuralism and morphological thinking*

Structuralism is historically linked to the constitution of the cultural sciences; they concretize the knowledge of human diversity formulated by the Enlightenment, that was already in Voltaire's *Essay on the Customs and the Spirit of the Nations*, Montesquieu's *The Spirit of the Laws,* Lessing's *Laocoon*, etc.

It is through the mediation of symbolic institutions that cultural diversity is expressed. To characterize it, we must not only compare these institutions, but also clarify their symbolic nature. Reflection on art has an eminent place here, for the works exemplify the creation of signs.

By the end of the 18th Century, the relationship between structuralism and morphological thought was widely established, from Goethe's morphological reflections to his references on the debate between Cuvier and Geoffroy Saint-Hilaire. All this was summarized by Humboldt, both philosophically and in terms of linguistic description.

Thus, the structural project is consubstantial with the cultural sciences: as comparatists, they want to constitute a general anthropology that takes into account all human, social, linguistic, artistic, legal and religious diversity, in short all those symbolic institutions that Cassirer called "symbolic forms".

The comparative undertaking demands a thrifty structuralism in its regulatory assumptions, which allows it to deal with complexity, instead of reducing it *a priori*, as the general grammars of the classical age did by universalist assumptions – now revived by Chomsky-ism. The axiomatic parsimony of structuralism opens up a reflection on the transposability and evolution of forms that allows its descriptive richness.

"Structuralism" was therefore not an episode in the history of the cultural sciences, but a conceptualization of their own historical and comparative methodology. However, two main conceptions can be distinguished without opposing each other: the first favors logical or logicized structures; the second favors morphologies and gives way to continuity and metamorphosis. They correspond to two preconceptions: one, of ontological tradition, favoring classifications and subsumptions of discrete or discretized objects; the second elaborates a praxeology to describe progressive individuations. They inspire two visions of invention and transmission that determine the evolution of cultures. This duality informs various conceptions of signs and semiotics and helps to clarify the place of this discipline within the cultural sciences.

However, morphological thinking reflects on natural forms, structural thinking on logical or even mathematical forms. René Thom tried to reconcile the two, through the realism of mathematicians: natural forms are also mathematical forms. Although Thom's theory remains somewhat disparate and programmatic as far as semiotics is concerned, the same postulate, of Pythagorean realism, was applied with consequence by Petitot in *Morphogenèse du sens* and *Les catastrophes de la parole*.

Morphological reflection based on the theory of catastrophes has favored fundamental, even transcendental processes and has therefore focused on predication, a crucial area of René Thom's reflection. However, it has neglected the anthropic areas that nevertheless determine the basins of attraction specific to teaching and that differentiate the forms of primary, secondary and ternary actancy. Finally, even reduced to narrative structures, textuality remained a distant horizon.

However, in relation to the Jakobsonian structuralism and its developments in Lévi-Strauss' and Greimas' work, the question of morphogenesis has made it possible to introduce two major fields:

– the continuum: Claude Zilberberg approached Greimas' work as early as 1966 from a Guillaumian point of view. However, the Guillaumian continuum (a form of tensivity) remains an Augustinian intentionalism, as evidenced by the very notion of *tension* (divided into protension and distension). The continuum also has a paradoxical status in semiotics, and particularly in linguistics: it exists only as an ideal substrate for variations differentiated by discretization;

– if one introduces continuous variations, one can favor continuist modeling, such as that of dynamic systems, as long as one favors singular points, favored as places of relevance. The singular points, where precisely the catastrophes that formalize the phenomena of hysteresis take place, are the places where the relevant features (phemes and semes) are located, since a form, through its transformations, is recognized by its singular points, such as Theaetetus' snub-nose.

Bringing together these two major fields, the theory of semiotic forms must account for their formation and evolution at all levels.

I.2.2. *Generic structures and genetic operations*

Classical 20th-Century structuralism sought invariants, elevating them to types and then describing empirical objects as variants. This approach took on a normative aspect: for example, in his famous *Morphology of the Folk Tale* (1927), Vladimir Propp chose as a model tale no. 300 from the collection of Aarne and Thompson, and then described the following tales as incomplete or deviant realizations of its structure. This method was theorized in 1929 by Bogatyrev and Jakobson in their article entitled *Die Folklore als eine besondere Form des Schaffens*.

Since the 1980s, Bronislava Kerbelyte has conducted a close critique of this folkloristics, insisting that the collection of oral traditions must take full account of the context of the practices, and that within them the problem of variation can only be addressed by a theory of creativity that supports a so-called "structural-semantic" methodology (Lemeškin 2018). This is a general problem: for example, jazz cannot be defined by a compilation of standards, which are known to all and whose potentialities the performers deploy.

If, however, structuralism, a general methodology in the cultural sciences, is defined as a search for *generic structures*, it remains to be completed by the search for *genetic operations,* in the sense of Simondon (2005): "the operation is what makes a structure appear or modifies a structure" (p. 270, author's translation), a search that we have designated by the term *operatics* (it seems preferable to the term *allagmatic* by which Simondon designated his own philosophy).

The duality between generic structures and genetic operations could undoubtedly be related to the opposition between logic-grammatical and rhetoric-hermeneutical problems, whose genealogy we have traced; at least the privilege granted to generic structures fits well with a consensual and, after all, traditional classificatory and typifying conception of science. On the other hand, the emphasis placed on genetic operations is better suited to the very purpose of the cultural sciences, since cultural objects are our works and we have to account for their creation and their evolution; this is why reflection on the arts has held such a place in the development of the cultural sciences, from Lessing to Goethe and from Cassirer to Panofsky. Their task of objectification thus differs radically from the natural and life sciences, as well as from the logical-formal sciences.

The duality between structures and operations can be related in linguistics to the Saussurian duality between language and speech, and in semiotics to the duality between instances and performances (Rastier 2011). In this duality, the side of operations generally dominates that of structures. Semiotic forms are the very place where the two sides are reconciled, since they are certainly characterized by specifying themselves in relation to their type, but on condition that their individuation operations are sufficiently completed or at least stabilized before their eventual de-individuation.

I.3. Logic, morphology and semiotic organon

In some of its aspects, logic can be used by the cultural sciences; for example, it has been used in linguistics for the study of argumentation. It could even be argued that logic is a (very) restricted part of semiotics.

Would not semiotics now be in a situation comparable to that of logic in the past, both as a discipline and as an organon for other disciplines? Locke, then Peirce and many others have identified logic and semiotics. This assimilation is based on the classical idea that thought is made up of operations on mental signs, which can be found from Port-Royal to Condillac to Destutt de Tracy. The relationship between mental signs and linguistic signs preoccupied the general grammars of the 17th and 18th Centuries, right up to the cognitive grammars of today – hence Chomsky's interest in *Cartesian linguistics*.

Historical and comparative linguistics has, however, disturbed the speculative relationship between universal mental signs and particular, variable and contingent linguistic signs, simply because it repatriates the signified in languages and delegates "mental content" to other disciplines, from logic to psychology. The problem of the relationship between empirical expressions and transcendent content is overcome, along with the traditional dualism between matter and spirit.

It is challenged by the duality proper to semiosis, defined as the pairing of content and expression: in terms of text, it characterizes textuality; in terms of the minimal syntagm, semiotization, understood as the pairing of a form of content and a form of expression. At the moment of this temporary taking of form, the sign becomes individuated and becomes a unique occurrence, a hapax. The closest and most repeated occurrences constitute a repertoire of stereotyped expressions and contents (phraseologies and doxa).

Classical logic operated on mental signs that were already discretized and that made up a repertoire, the same for grammar for linguistic signs; this allowed term-to-term pairings, both in logic and in grammar. On the other hand, as the study of texts attests, linguistics deals with forms that escape logic because of their continuous nature and, because of their constant transformations, call for a morphology, even a *morpho-logic*.

Morphological reflection has been present since the beginnings of linguistics, notably in Humboldt's work, who was inspired by Goethe's reflections, and its importance was recalled by Cassirer in his final article "Structuralism in modern linguistics". Moreover, the "semiotic square" (according to Greimas and Rastier 1968) gave concrete expression to the general idea that there was a need for a non-logical theory of the operations of constitution and evolution of semiotic forms. Although it called for morphological (and not logical) developments that were slow in coming because of the importance of the logic-grammatical problem, it was unfortunately considered a model (in every sense of the word) and ended up being iconized in an acritical manner.

A morphological reconception of structure is now necessary. It was anticipated by a morphosemiotics such as Pottier's, elaborated since the early 1960s, notably in his model of Latin prepositions (Pottier 1962) and up to his *Sémantique illustrée* (Pottier 2018), and of course Petitot's contributions on the morphogenesis of meaning. Morphosemiotics proposes new answers to the dual problem of the transposability of structures and their evolutionary regimes, notably by mobilizing the theory of dynamic systems (Rastier 1999; Piotrowski 2017).

In short, logic models represent only a very small number of mathematical models; more powerful and diversified, by determining the discrete as a "special case" of the continuous, they make it possible to pose both the perceptual and cognitive problem of discretization. Group theory, dynamic systems theory, topology and catastrophe theory have been used in various ways to deal with problems of semiotic morphology. Corpus linguistics has also exploited probabilistic models to identify forms, both by internal correlations between their singular points, favored by "classical" structuralism, and by outline segments, favored by the morphological approach. In addition, it seeks to characterize shapes by their reciprocal contrasts.

These models escape the all-or-nothing principle and retain proximities or distances. Beyond this, it is the need for corpus semiotics that is now essential, not only for the criticism of "data", but also for the reasoned instrumentation of research (Rastier 2018b).

I.4. Semiosis

The signs that surround us are not given to us, because we reproduce them to interpret them each time they occur. Present in all our actions and no doubt in what we confusingly call our thoughts, they are both the products of a collective work and the material of our speech and writings.

Traditional in philosophy, the separation between language and thought remains accepted in cognitive linguistics. However, this separation is relativized as soon as one renounces ordinary mentalism in semantics: content and expression become inseparable.

Rather than an ontology, the signs are based on a theory of action or *praxeology*. This weakens logicist conceptions of language as a "code" and leads to considering it as a system constantly modified by its use and worked by historical dynamics (Rastier 2018a).

Like any metastable result of an individuation process, the sign corresponds to a local minimum of energy, but its creation has provided it with an activating energy and thus maintains the tension that will allow further transformations. Any semiotic quantity is indeed metastable in a particular sense: since it is purely differential, it is sufficient to modify its paradigm, its context or its corpus for it to change with the relations that constitute it. Thus, it never reaches a stable state, since it is always part of an evolving system and corpus. Its repetition remains impossible, because a sign is only perceived as such in a context of production or interpretation, i.e. of reproduction, understood as re-creation.

A distinction must be made here between the random variations that affect any occurrence considered as a hapax[2] and the convergent variations that come from a project and contribute to a new taking of form, hence the need to elaborate the theory of transformations of semantic and expressive forms and backgrounds.

For this, it is no longer a question of continuing to oppose sight and hearing, but of differentiating between a static ontological conception that has taken as its fundamental metaphor the persistence of the object perceived by the vision and a dynamic praxeology that draws on the fluency of phonation and the variety of enunciative gestures.

By conceiving language in the image of another semiotic form, writing and especially alphabetic writing, grammar has conformed it to a partly compatible but heterogeneous model. The discretion of the characters (and then of their groupings, when the *scriptio continua*, which left no space between words, came to an end) discourages us from considering continuous transitions and the dynamics specific to primordial phenomena such as prosody, which transcends and controls syntactic divisions that are, moreover, poorly adapted to the spoken word. Other phenomena, of isotopy and isophony, semantic and phonetic contrasts, and vocal and emotional tonalities, seem to be prominent everywhere and ensure the relative continuity and coherence of the enunciative and interpretative courses of action.

Also, more unpredictable than strictly chaotic, the two flows of meaning and expression cannot be completely reconstructed by their points of coincidence, since the simple examination of signs is not sufficient to characterize the semiosis movements from which they result. We can certainly evoke a mysterious enunciation, a cognitive or other type of subject that has become the *deus ex machina* of the syntactic machinery that it providentially sets in motion; but the project of a statement is always collective, if only because of the dimension of the address. The articulation of the social and the individual is thus played out, in an

2 Saussure gave the illustrious example of repetition *Gentlemen! Gentlemen!*, where each occurrence differs in position, intonation and so on.

action theory, a praxeology incompatible with the ontology that settles the logic-grammatical issue.

In relation to the very structure of the sign as a cultural object, the duality between the two conceptions of semiotics, restricted and extended, is reflected in the duplication between the content and scope of the sign. Since the meaning of the sign governs the textual meaning, its content can only be completely determined by its scope. At the same time, semiology as a description of sign systems depends on the description of the social institutions that constitute them. Moreover, they do not reflect them, because they are part of them.

As Saussure saw it well in forming the project of a general semiotics, a renewed conception of language and of the specific objectivity of linguistic signs has great consequences for the understanding and description of all other kinds of signs as well as of all cultural systems.

I.5. The question of interdisciplinarity

The repeated invocation of interdisciplinarity can remain incantatory if it merely translates, betrays and compensates for the fragmentation of disciplines on an illusory scale, conducted in the name of an acritical hyperspecialization.

Interdisciplinarity remains a problematic notion, or even a mere slogan, as long as we remain with uninformed generalities, without distinguishing for each discipline its objective, the specificity of its object and its methods of objectification. Moreover, in certain humanities circles influenced by deconstructive philosophy, the very notion of discipline is eroded in the face of a discourse that is repugnant to objectification and whose very length attests to the fact that no criteria allow it to conclude. The very notion of discipline then appears disciplinary, in the repressive sense of the term.

Interdisciplinarity remains a resolution as long as there is no common language or "translation" procedure between disciplines. For structuralism, it was philosophy, and the masterful philosophical reconstruction of the sciences by Cassirer and Simondon that was intensified by the philosophical reflections of scientists such as Lévi-Strauss (who, moreover, "defected" from philosophy, for which he was originally destined). This is why, since the end of the 1960s, deconstruction has sought to shatter the common language of the philosophy of science through an aporetic discourse that has become anecdotal, but largely dominant in certain areas of the humanities. Forty years after Heidegger's attack on Cassirer, Derrida's attack on Saussure amounted to a mock trial based on a largely apocryphal text. It had a symbolic meaning: to dissociate the cultural sciences from the "thought" that

Heidegger allowed himself and perpetuated his rejection of principle of all science and made it impossible for them to self-reflect in philosophy.

Another common language was that of "formal philosophy", as Richard Montague's editor put it. Russell's great current of logistics, to which Whitehead gave a largely Platonic interpretation, was reminiscent of the age-old cousinhood of philosophy and mathematics. In the early 1930s, the formalization program underwent a materialistic inflection with the Vienna Circle. The project for the unification of sciences, subsequently formulated by Morris and Carnap in the *Encyclopaedia of Unified Sciences*, was marked both by its ambition to unify all sciences and by scientific criteria that effectively excluded most of the cultural sciences.

Popper went on to formulate criteria of scientificity that are well suited to the experimental sciences but hardly appropriate to most cultural sciences; when they are strictly applied, as they have been by influential decision-makers, they lead to their marginalization. The image of science conveyed by the Popperian criteria has been called "Galilean" by Milner in his praise of Chomsky; his *Introduction à une science du langage* started linguistics relative the Chomskian project and remained silent on the century and a half that preceded it.

Based on the image of Newtonian physics, this conception of science hypostasizes the principle of causality and sums up knowledge in the form of laws. It has been, if not made obselete, at least overtaken by physics and mathematics (especially non-Euclidean geometries) since the turn of the 1880s and 1920s. In the cultural sciences, causes give way to conditions, and to correlations between conditions; the statement of laws to the conjectures of probabilistic reasoning; finally, regularities are no longer absolute but leave room for singular events.

The consensus on what a theory is then conditions two types of interdisciplinarity:

– the first, which has remained unnoticed or is perceived as paradoxical, is *internal interdisciplinarity*: because of the growing disciplinarity, which leads, for example, to the view that linguistics is giving way to language sciences, or that morphology is a discipline other than syntax, it is important to reaffirm the hard-won unity of disciplinary fields. Saussure masterfully formulated this in this definition: "Semiology = morphology, grammar, syntax, synonymy, rhetoric, stylistics, lexicology, etc., *all being inseparable*" (Saussure 2002, p. 45, author's translation);

– *external interdisciplinarity* has remained a keyword for decision-makers. Paradoxically, it works better between scientific departments than within the same department: however, an archaeologist collaborates without a fuss with a palynologist, but between social sciences there remain squabbles over academic and sometimes scientific boundaries.

Let us not postulate a supra-theoretical unity, but let us count on a common desire to recognize the complexity of cultural facts and on a common project of objectification. Although each one differs in its methods and validation formulas, they remain united by the principle of reality and the concept of scientific truth.

In several disciplines of the humanities, especially in the identity-based currents of cultural studies, where theory is reduced to a series of proper names, dialogue with the sciences is abandoned and a program of "critical intervention" is substituted for the knowledge project. However, a scientific result is the result of a critical activity, which is necessary for scientific deontology. Once established, it is not critical in itself, but is nonetheless de facto critical of the prejudices it dispels, so that scientific truth can be emancipatory, even revolutionary.

I.6. Mutual aid between sciences

As a discipline, semiotics has its place among the cultural sciences, particularly in order to differentiate and compare the different "symbolic forms" or semiotic institutions such as myth, art, law or language.

As an organon, semiotics can propose conceptualizations, or even types of formalizations that each discipline can take up in order to make them operative within itself for its own problems.

However, the situation of semiotics does not give it the status of an exception: all disciplines participate in circles of cooperation, where each can assume an ancillary role with respect to the others, while helping each other in return for their collaboration. For example, logic and more generally mathematics can guide semiotics in the elaboration of its models and its instrumentation, as evidenced by interrogation and exploration software in corpus linguistics, or even in iconology or musicology.

Disciplines meet on problems arising from their applications. Thus, cultural sciences such as archaeology, history and ethnography cooperate with disciplines in the life sciences, such as palynology, and make use of techniques such as spectrography for their dating and restitution of disappeared contexts.

Finally, it should be recalled that semiotics also extends to the life sciences, as evidenced by the rise of zoosemiotics; however, this would call for its own development, which is beyond the scope of this introduction.

Let us return to identity concerns that reject culture for the benefit of cultures. For a semiotics of cultures, it is impossible to project categories *a priori* on its objects. For example, in linguistics, we make no difference in principle between minority and majority languages, and even between living and dead languages. The same is true of cultures: to know the ever-local and evolving specificities of a culture, one must compare it with others, if not all the others. A culture can only be understood within the corpus constituted by other cultures, to which it is linked by a cluster of similarities and differences, borrowings and innovations of its own.

While deconstruction pleads for the reconquest of alienated identities, the cultural sciences, through the methodology of their comparative project, break with all essentialization. Identity is only a metaphysical delusion: it can only be affirmed tautologically, but never demonstrated. If, therefore, the cultural sciences have a political impact, it is by ultimately bringing identity discourses back to their own inanity, to enable a better understanding of the relationship between the individual and the particular, on the one hand, and the general and universal on the other.

I.7. References

Greimas, A.-J., Rastier, F. (1968). The interaction of semiotic constraints. *Yale French Studies*, 41, 86–105.

Lemeškin, I. (2018). Étude structurale du folklore [Online]. *Texto ! Textes et Cultures, Revue de l'Institut Ferdinand de Saussure*, XXIII(4). Available at: http://www.revue-texto.net/index.php?id=4161 [Accessed 19 August 2019].

Piotrowski, D. (2017). *Morphogenesis of the Sign*. Springer, Berlin.

Pottier, B. (1962). *Systématique des éléments de relation*. Klincksieck, Paris.

Pottier, B. (2018). *Sémantique illustrée*. Académie des Inscriptions et Belles Lettres, Paris.

Rastier, F. (1999). Cognitive semantics and diachrony. In *Historical Semantics and Cognition*, Blank, A., Koch, P. (eds). Mouton de Gruyter, Berlin, 109–144.

Rastier, F. (2011). *La mesure et le grain – Sémantique de corpus*. Champion, Paris.

Rastier, F. (2015). *Saussure au futur*. Les Belles Lettres/Encre marine, Paris.

Rastier, F. (2018a). *Faire sens. De la cognition à la culture*. Classiques Garnier, Paris.

Rastier, F. (2018b). Computer-assisted interpretation of semiotic corpora. In *Quantitative Semiotic Analysis*, Compagno, D. (ed.). Springer, Berlin, 123–139.

de Saussure, F. (2002). *Écrits de linguistique générale*. Gallimard, Paris.

Simondon, G. (2005). *L'individuation à la lumière des notions de forme et d'information*. Jérôme Millon, Grenoble.

PART 1

Semiotic Foundations of the Cultural Sciences

1

Cassirer and Symbolic Forms

This chapter aims to discuss how Cassirer's reflection on the internal diversification of meaning accounts for the inner transformation process of science. In order to do so, it will be necessary to place Cassirer's approach in the Kantian tradition before showing how and why he moves away from it. The relevance of his point of view today will be shown by means of an example.

1.1. Unity and diversity of modes of objectification

1.1.1. *Modes of objectification in the transcendental tradition*

Cassirer is generally credited with having succeeded in legitimizing the idea of an intrinsic variation in modes of objectification, which he called "symbolic forms" – his main contribution to the Kantian tradition he claimed to be part of. Because of him, this tradition would thus have passed from a univocal idea of objectification as realized in scientific knowledge to a plurality of these modes in which science would no longer play the leading role but only a part, being on an equal footing with other modes. One then generally presents things by exhibiting, in a very empirical way, several lists of these modes that Cassirer gave on occasion: to the first three well-known modes of language, mythical thought and scientific knowledge as they appear in the three volumes of his *Philosophy of Symbolic Forms* would have been added later on as well as various others.

However, a quick glance at Kant's work shows, without even going into detail, that this representation of Cassirer's contribution is largely at fault. Kant did indeed come up against the problem of the plurality of modes of objectification, since one was rationally constrained to superimpose a supra-sensitive reality to sensitive

Chapter written by Jean LASSÈGUE.

reality – i.e. the causally determinable nature as described in *Critique of Pure Reason* – by postulating a free causality for the purpose assigned to actions as described in *Critique of Practical Reason* [1]. This simple reminder suffices to remove the authorship of the problem of the variety of modes of objectification from Cassirer in the tradition of transcendental philosophy, and requires us to re-examine the notion of what the symbolic form represents in a tradition that had posed the problem long before him.

Kant's answer to the question of the plurality of modes of objectification was to distinguish three different cases, each of which was the subject of his own elaboration in one of the three *Critiques*: the transcendental, which Kant described as "narrow", delimiting the domain of knowledge in the strict sense; the transcendental "in the broad sense", which he also called "metaphysical" and which delimited the domain of morality; and finally what we must call, even if the term is not of his own making, "the reflexive", which delimits the domain of art and biology conceived as a finality without purpose. It is the difference between the pure and the empirical that makes it necessary to distinguish these three cases: between the two diametrically opposed cases of scientific knowledge susceptible to phenomenal experience and moral judgment that is radically devoid of it, the reflexive is phenomenalized by the form of objects with circular causality. From this point of view, the transcendental relates to pure concepts, unlike the other two, which have a different relationship with the empirical datum, since certain characteristics of the concepts they analyze depend on it in the two ways just mentioned. Thus, in particular, one must distinguish the pure that is *independent* of experience from the *a priori* that is only *prior* to it. For example, while physical science is pure and *a priori*, its analysis is therefore part of the transcendental in the strict sense, morality is *a priori* without being pure because the concept of will conceived as a faculty of desire is not[2] and its analysis is therefore part of the metaphysical mode, in the sense

1 "This law must give the sensible world, considered as a sensible nature (as far as reasonable beings are concerned), the form of an intelligible world, that is to say, of a supra-sensible nature, without attacking its mechanism" (Kant 1848, p. 196).

2 "The third Criticism repeats it: the concept of will as the faculty of desire, always given empirically, escapes the transcendental. Will does not belong to transcendental philosophy, but to the metaphysics of morals. 'A transcendental principle is a principle by which is represented the *a priori* universal condition under which only things can become objects of our general knowledge. On the other hand, a principle is called metaphysics when it represents the only a *priori* condition under which objects, whose concept can be given empirically, can continue to be determined *a priori*' (*Critique of Judgment,* second introduction, section V, Ak V, p. 181; II, p. 935). Kant gives two examples of this transcendental/metaphysical distinction. The first distinguishes between knowledge of bodies as substances, a transcendental principle in the strict sense, and knowledge of bodies as modifiable substances, a metaphysical or transcendental principle in the broad sense" (Lequan 2007, p. 7).

that this term takes on in Kant's work, for example in *Groundwork of the Metaphysic of Morals*. This distinction between pure and *a priori* has been the subject of revisits in the Kantian tradition.

It was in fact the neo-Kantian philosophers of the Marburg School, of which Cassirer was a student, who observed that the transformations of science during the 19th Century (projective geometry, kinetic theory of gases), which took place after Kant's death in 1804, required a refoundation of the transcendental project: it was necessary to take into consideration the advancement of science, essentially in mathematics and physics, and more precisely to justify this advancement from a transcendental point of view. This was a legitimate observation from the Kantian point of view: Kant himself would certainly not have denied it, as he had clearly specified that the *Critique of Pure Reason* was a "propaedeutic" to transcendental philosophy and not this philosophy itself, which was still to come, insofar as the *Critique of Pure Reason* did not expose the content of science in an exhaustive manner but only its rudimentary principles.

1.1.2. *The geometric objectification crisis*

If the progress of science really were to *contradict* the transcendental principles as Kant had formulated them, what would become of the Kantian tripartition between the transcendental, metaphysical and reflexive modes of objectification? For if science was likely to evolve not only in its content but also in its principles, this would automatically place it on the side of metaphysics and no longer on the side of the transcendental, *since this evolution would present itself as an empirical fact*, contingent on a particular history and no longer as a pure concept. If this were indeed the case, the transcendental point of view would have to be abandoned in order to allow the progress of science, which some philosophers, in particular those of the Vienna Circle, did not hesitate to do.

Yet it was precisely this situation of *vacancy of principles* that Cassirer was confronted with when he began his studies of philosophy: the question of the transcendental ideality of space had been at the heart of the epistemological debate since the discovery that space not only had the Euclidean properties attributed to it by the transcendental aesthetics of the *Critique of Pure Reason*, but that other geometries were just as legitimate, even if they did not fit into the Kantian framework. The problem of the advancement of science consists not only of the new appearance of phenomena but of the appearance of *new phenomena* whose intelligibility requires a renewal of principles. More precisely, what to do philosophically because of the advent of non-Euclidean geometries at a contingent moment in time, in the second half of the 19th Century? And what are the consequences for our conception of the transcendental and other modes of objectification in general?

If the transcendental is to retain its value philosophically, we must succeed in showing, as Cassirer did, that *it is the transcendental that demands that the possible evolution of the principles of science be integrated into itself*. It is therefore this new interpretation of the notion of the transcendental that justifies the variety of modes of objectification. The case of geometry is exemplary from this point of view in that it now deploys within itself a variety of modes of objectification (Euclidean and non-Euclidean geometries) that can be exported elsewhere (in relativistic physics in particular). Philosophically, therefore, Cassirer's gesture consists of recognizing a double movement at the heart of the transcendental: on the one hand, an *internalization* of the variation of the modes of objectification – the intelligibility of this variation being not limited to the relation to experience anymore; on the other hand, a possible *exportation* of these modes to domains different from where they emerged, including towards experience as is the case of non-Euclidean geometry in relativistic physics. From this point of view, Cassirer's gesture in philosophy was analogous to Einstein's in physics: just as Riemannian geometry is exported into a new framework of intelligibility in physics so are the symbolic forms, the validity of which expand to domains that are not supposed to be rational like myth or language. As Cassirer pointed out:

> Each of the original directions of knowledge, each interpretation, which it makes of phenomena to combine them into the unity of a theoretical connection or into a definite unity of meaning, involves a special understanding and formulation of the concept of reality. There result here not only the characteristic differences of meaning in the objects of science, the distinction of the "mathematical" object from the "physical" object, the "physical" from the "chemical", the "chemical" from the "biological", but there occur also, over against the whole of theoretical scientific knowledge, other forms and meanings of independent type and laws, such as the ethical, the aesthetic "form". It appears as the task of a truly universal criticism of knowledge not to level this manifold, this wealth and variety of forms of knowledge and understanding of the world and compress them into a purely abstract unity, but to leave them standing as such. Only when we resist the temptation to compress the totality of forms, which here result, into an ultimate metaphysical unity, into the unity and simplicity of an absolute "world ground" and to deduce it from the latter, do we grasp its true concrete import and fullness. No individual form can indeed claim to grasp absolute "reality" as such and to give it complete and adequate expression. […] It is the task of systematic philosophy, which extends far beyond the theory of knowledge, to free

> the idea of the world from this one-sidedness. (Cassirer 1923, pp. 446–447)

These various symbolic forms are not watertight and can be found in forms other than their original one, they are *exportable*: language and art are found in scientific knowledge as well as mythology in language or law. Everything is therefore a matter of *accentuation* according to what the symbolic form of the situation particularly emphasizes but without ever breaking with the unity of meaning.

For Cassirer, therefore, the transcendental is no longer the analysis of the pure conditions of knowledge fixed once and for all within a homogeneous space but the internal transformation of the principles of knowledge into particular forms of objectification and their possible diffusion in various fields from the contingency of historical situations. Cassirer underlined this:

> The "fact" of science is and stays of course, according to its own nature, a fact which develops historically. While in Kant this insight has not yet been recognized unambiguously, while categories, as far as their number and content are concerned, can still appear in him as ready-made "fundamental concepts of the understanding", the modern enhancement of critical and idealistic logic has brought full clarity on this point. The forms of judgement only signify unifying and living motifs of thinking which traverse through the whole diversity of its specific forms and mobilize themselves in creating and expressing categories that are always new. The more these variations are proved to be rich and flexible, the more they confirm the specificity and originality of the logical function from which they emerge. (Cassirer 1906, Band 1, ECW 2, pp. 14–15, author's translation)

From then on, it is no longer, as in Kant's work, the pure/empirical dichotomy that makes it possible to distinguish between the different modes of objectification, since the "pure" itself possesses a possible internal diversification. Only the notion of *meaning* makes it possible to account for this diversification, which is no longer attached to the pure/empirical difference and which manifests the diffracted unity of the forms of objectification. Each mode of objectification thus institutes a particular way of producing meaning by giving it a form, and meaning is a local product of the never entirely stabilized relationship between these forms since it is the contingent result of their harmonics. Cassirer points out in this regard:

> [...] each particular category which we single out and place in relief against the others, can only be interpreted and judged as a single factor which may develop very different concrete configurations

> according to the relations into which it enters with other factors. (Cassirer 1955, p. 269)

The question that remains then is how to conceive of the reciprocal relationship between these "themes" or "forms" in light of the fact that we can no longer rely on a uniform space underlying any phenomenal manifestation to account for their possible reciprocal relationships, as was still the case for space in the transcendental aesthetics of the *Critique of Pure Reason*. Only the notion of meaning, in Cassirer's eyes, allows us to understand these relationships because it does play the role of a condition of transcendental possibility of space:

> Space does not possess an absolutely given, final, and fixed structure; rather, it acquires this structure only by virtue of the general coherence of meaning within which its very construction is accomplished. The function of meaning is the primary and determining one; the structure of space is a secondary and dependent element. (Cassirer 2013, p. 325)

The notion of meaning thus mixes the registers of the pure and the empirical – which would seem to Kant to be a contradiction in terms – and a philosophy focused on the study of modes of objectification must analyze how meaning varies and differentiates itself progressively in particular semiotic situations.

1.2. The harmonics of forms: internalization and exportation

Getting away from the Kantian question of the pure/empirical distinction in order to account for the modes of objectification therefore involves internalizing the conditions of variation specific to a field, as Cassirer's case of geometry has historically shown. The mathematical concept of the transformation group is therefore in a position to be exported to other fields in order to contribute to the internalization of the conditions specific to the variation in the fields in question.

1.2.1. *Interdisciplinarity of the transformation group concept*

Cassirer showed that the concept of the transformative group is found across the knowledge spectrum, whether in the natural sciences (mathematics, physics and biology) or in the humanities and social sciences (linguistics and psychology). It is important to note, however, that the interdisciplinary use of this mathematical concept does not mean that it can be used indiscriminately in the disciplines in question, which would be tantamount to presupposing a common underlying space, a kind of pre-Einsteinian ether, if one wanted to spin the metaphor of the relationship between Cassirer and Einstein. Thus, rather than sticking to a description of its relevance in all the disciplines listed – which would precisely risk giving the impression of a homogeneity in the conditions of its application – we will stick to a single case, that of the difference between geometry and perception, because it shows how Cassirer distanced himself from the Kantian point of view *stricto sensu*.

Kant would have identified the difference between the two fields as typically pertaining to the difference between the pure and the empirical and he would have thought of the Euclidean space as the unique place of their possible relationship. On the contrary, Cassirer needed to show that the difference between the mathematical and the perceptive was fully recognized and that it was not the postulation of a common homogeneous space that made it possible to relate them, but rather specific conditions for the use of the concept of a transformation group in both domains.

Cassirer insisted on the absolutely heterogeneous character of the mathematical and perceptive domains:

> The first question I would like to ask is this: is it by pure chance that, in the pure and simple relationship of the psychological fact, a concept belonging to the theory of groups slips in? One might be disposed to see, in the use of this concept, only a simple equivocation or at most a metaphor. This use should in no way lead to pursuing the dream of a "mathematical psychology", as presented in a purely speculative outline by Herbart, for example. The "rigor" of mathematical concepts comes from the fact that they are attached to a certain circle of ideas. They cannot, without suffering logical damage, be employed outside that circle; they cannot simply be "transposed" into other areas. (Cassirer 1938, p. 382, author's translation).

There is therefore, strictly speaking, no "exportation" of the group concept from the "pure" to the "empirical" – the very scheme of "abstraction" would be too empirical – there are, on the other hand, *different uses* of the concept of the transformation group depending on whether it is used in scientific theories or in the theory of perception. Indeed, in the case of its use in the sciences, whether "exact" or "human", the use of the group concept aims at transforming what already has the status of a *theory* by varying what is meant by "object" within it. For example, in the geometric case, the use of the notion of the transformation group increases the possibilities of constructing geometric objects and their internal organization according to the type of group chosen. In the case of perception, on the other hand, the notion of the transformation group makes it possible to describe a certain number of laws of organization of the form such as they were updated by the *Gestalt theory* laws that make it possible to account for the perceptive constancy in spite of changing physical conditions[3]. The concept of a transformation group is therefore very relevant, but the relationship with the object is profoundly different from the geometric case in that it does not aim to study the possible variations of geometric objectification *for their own sake*, but only to choose among them the one that seems to best account for a field of perception that is itself variable. The pure/empirical difference is thus replaced here by the more or less pronounced emphasis on the exploration for itself of the universe of groups or on the choice among them of a particular group to account for the facts of perception. What was considered "empirical", i.e. the "given" that the subject renounces to construct, is thus replaced here by a choice that is part of the subject's semiotic activity, the expectations of which have yet to be described.

1.2.2. *Beyond the transformation group*

It would, however, misrepresent Cassirer's philosophy to suggest that the concept of the transformation group plays a transcendental foundational role in it. The concept of the transformation group is not necessary to Cassirer's philosophy: it only makes it possible to represent a state of knowledge at a certain time, the time when Cassirer was intellectually active, and his reflection on its use was not intended to make it the equivalent of a category fixed once and for all. It is not the potential mathematization that it contains that makes it a foundation, because mathematics no longer plays the role it played in Cassirer's work as it did in Kant's work: mathematization is no longer the supreme goal of all knowledge, but only the possible objectification of meaning in particular situations that are not intended to form a homogeneous structure as philosophy has tried to conceive it from the classical project of *mathesis universalis*. In fact, the notion of a transformation group is only one of the possible modalities of the permanent revival of meaning conceived as a *potential for transformation*. This revival is Cassirer's minimal definition of the transcendental and can take forms other than the concept of the transformation group, even in mathematics.

This is the case of Cassirer's example related to the principle of counting in Sotho, a Bantu language, in which he noted that the counting procedure relies on the body parts, since it is based on counting fingers and then toes. Contrary to a number of philosophers of his time[4] who drew racist conclusions about the "primitive

3 "The 'true' color, the 'true' shape, the 'true' size of an object in no way coincide with what can be given to us in an isolated impression, nor can they be composed of these impressions. The function of memory and the call of the processes of reproduction are not enough to create it either. The constitutive factor is elsewhere; it lies in the possibility of the formation of certain invariants" (Cassirer 1938, p. 413, author's translation).

4 Steinthal in particular who Cassirer quotes (Cassirer 1972, p. 190).

character" of this population, Cassirer showed that the procedure in question is based on the need for *a sequence in an identical order*. Hence, the particular name of certain numbers in Sotho, such as the number 6, which means "jump" (implied: "to the other hand")[5]. Counting in Sotho thus reveals the need for the mathematical rule of well ordering and thus paves the way for the possibility of constructing the concept, an indispensable prerequisite for the proper mathematical definition of the number as Cassirer conceived it from the work of Dedekind, his contemporary[6].

Of course, this revival of meaning is not limited to the mathematical field and one could cite other examples in other fields that have nothing to do with the concept of the transformation group. This is, for example, the case for his conception of law in the polemic that put him in opposition with the Swedish philosopher Hägerström. For Cassirer, Hägerström was a typical example of the dangers of the philosopher's exercise outside the transcendental tradition in that he was rationalist in the philosophy of knowledge but irrationalist in moral philosophy, because he was incapable of understanding the necessity of the variety of modes of objectification. Thus, Hägeström, being a rationalist, could not admit, for example, that Roman law may have mythical foundations: if Roman law had mythical origins, then the whole rational theory of later law collapsed. For Cassirer, on the contrary, the mythical origins of law did not take anything away from its later rational aspect: it was precisely the effort of rationalization of which the progressive theorization of law was capable, which would gradually rid it of what would become, once the meaning had been transformed, a mere mythological trace:

> But here too the question is raised whether the insight into the genetic beginnings of culture can give us an insight into its various functions [...]; the question is not what these functions whether language, art, law were originally but what they became by virtue of this transformation of meaning. (Cassirer 1939, Band 21, s. 86–87, author's translation)

5 "Thus the numerals do not so much designate objective attributes or relations of objects, as embody certain directives for the bodily gesture of counting. They are terms and indices for positions of the hands or fingers, and are often couched in the imperative form of the verb. In Sotho, for example, the word for 'five' means literally 'complete the hand', that for 'six' means 'jump', i.e., jump to the other hand. [...] By this method, the motions of arranging the objects are coordinated with certain bodily motions which are conceived as running in a certain order" (Cassirer 1955, p. 230).

6 "The ordinal number theory thus represents in fact the minimum requirement which no logical derivation of the notion of number can renounce, whereas the consideration of equivalent classes, if it is of the greatest importance for the applications of this notion, is of no interest for its original content" (Cassirer 1977, p. 69; Cassirer 1998, volume 6, p. 54, author's translation).

It is therefore the potential for principles of knowledge to be transformed, whether it takes a mathematical form as in the case of the transformation group concept or any other form of transformation of meaning, that makes it possible to account both for the institution of a new form at the crossroads of other forms and for its possible progressive diffusion through all fields of knowledge. An example that is not Cassirer's, to conclude, can show the fruitfulness of this approach today.

1.3. From the social sciences to the natural sciences and back again: the example of statistics

The philosopher of science Ian Hacking noted long ago in *The Taming of Chance* (Hacking 1990) that statistics first appeared in the late 18th Century in the social sciences (mainly in censuses and birth and death lists in France, England and Prussia) and that this was linked to a governmental method typical of the modern state for which a statistical service was indispensable to the administration of its public policies. It was only later that the concept of statistics migrated from the social sciences to the natural sciences, to the point of investing in the very heart of atomic physics with quantum mechanics in the first years of the 20th Century. This is a very good example of a kind of revenge of the so-called "soft" sciences on the so-called "hard" sciences, which shows that concept transfers can head in the direction of the natural sciences toward the social sciences, but that the opposite direction is also evidenced. In short, while there is a tendency to see how the scientific viewpoint invests in fields that are deemed not to be so, the opposite is also true, as the case of statistics shows. There are two things I think are important to emphasize in this example. First, the questioning of the precedence, implicit but very much alive, of the natural sciences over any other mode of knowledge. But we cannot simply stop at this war of disciplines, however justified it may be in this case, because it concerns primarily conflicts of institution and not the content of knowledge itself. Second, Hacking's work has placed little emphasis on the important point that the social sciences are transformed in *return* once the concept of statistics has passed through the natural sciences. By contrast, this is one of the strong points made clear by mathematician and philosopher Olivier Rey in his book, *Quand le monde s'est fait nombre* (Rey 2017) (meaning: When the world became number), which takes up the issue very carefully and shows that by transiting through the natural sciences, the concept of statistics allows a *new look at the nature of society*. But to get to this point, which is crucial for the place that a philosophy of science must occupy in a philosophy of culture, we need to go through a little history of statistics.

1.3.1. *The internal historical transformation of the statistical paradigm*

My hunch is to interpret Rey's point of view in the Cassirerian way I have developed above, that is by defending the idea that, in the social sciences, the question should not be to "do things like science" (implied: of nature) – because this would imply the complete homogeneity of the field of knowledge that Cassirer had renounced – but by letting the concept of statistics span over various forms of knowledge (in this particular case, natural sciences), to get a better idea of what is at stake in yet other forms of knowledge (in this particular case, social sciences). This process is made possible without supposing an underlying unity of science that would only be cheaply cemented and would suggest that the validity of concepts is immediately universal regardless of the field in which they are used.

There are four steps to report on the development of a statistical perspective. In the first step, Quételet appears as a founder: a mathematician first interested in astronomical questions, he then studied a series of measurements of the human body – height, weight, etc. – which were used to determine the size of the human body. This gave rise to the concept of the "average Man"[7]. The measurements in question drew a bell-shaped curve whose central value defined the "average Man" who was also an *ideal* Man, a standard of humanity for a given population and which was absolutely unchangeable, even if it were constructed statistically. Deviations were due to *errors* in measurement and were not relevant in themselves. The idea of science therefore remained a complete determination: astronomy, where small perturbations are considered negligible, therefore still plays the role of a paradigm in Quételet's view. But this full determination is not accessible to human knowledge, as Laplace notes in his *A Philosophical Essay on Probabilities*. It is precisely this idea that will gradually be called into question in the gradual transfer of the concept of statistics to the natural sciences. Without going into historical details that are not of direct interest here, let us note the three subsequent stages.

In the second step, the physicist Maxwell explicitly compared the case of gases to the case of societies composed of a host of individuals. He noted that the speeds and directions of molecules were not known with sufficient precision to be able to stick to the deterministic laws of Newtonian mechanics. The limitations of our knowledge (in the way described by Laplace) forced one to resort to new laws of a statistical nature. But Maxwell was truly innovative when he showed that the perturbations, that is to say all parts of the normal curve, were worthy of interest: the variations were not inessential; on the contrary, they were the very *heart* of the phenomenon. Such a remark therefore already implied a mental conversion because

7 "[...] the measurements of many objects of the same nature produced the same curve as the many measurements, with varying degrees of error, of the same object" (Rey 2017, p. 179, author's translation).

the statistical method was no longer considered as a last resort[8]. Here, there was a first relativization (and perhaps not an "abandonment", as Maxwell's text says) of the deterministic model (Newtonian–Laplacian) in physics that aimed at integrating in the framework of physics an *internal* differentiation according to the scales of measurement by recognizing a *local* mathematical relevance to statistics, which was no longer immediately multiscale. The problem that arose was obviously the articulation of these levels since the microscopic level was constituted by the deterministic equations of Newtonian dynamics while the macroscopic level was constituted statistically.

It is Bolzmann who, in the third step, solved the question by using *statistical calculation* itself, allowing for an all-encompassing approach to the question by showing that a distinction must be made between observables (macrostates) and atoms (microstates) and that certain configurations of microstates are more frequent than others because they necessarily occur more often (as shown by the bell curve), giving rise to more frequent macrostates. It is therefore possible, by varying the scales, to articulate deterministic and statistical laws by situating the global framework of knowledge of physics beyond this distinction.

It then became possible, in the fourth step, to also consider determinism as an *approximation* of the statistical framework, an approximation that could finally be dispensed with, as Exner did, who proposed a unified vision of the knowledge of nature, assuming that determinism did not apply at the microscopic level either. The theoretical framework had therefore completely changed: it was by relying on the *impossibility* of an exhaustive analysis of the elements studied that a determination became conceivable, contrary to the deterministic attitude, which consisted of establishing an equivalence between the possibility of science and the exhaustiveness of the elements it studied. This was a complete reversal of the Newtonian–Laplacian paradigm, since science *derives its fruitfulness from the impossibility of grasping a situation* and it is an intrinsic lack of knowledge that allows the elaboration of a certain form of knowledge: science aims to make explicit the *local* conditions of possibility of a certain regime of knowledge[9].

8 "The statistical method [...] which, in the present state of our knowledge, is the only one available to us for studying the properties of material bodies, implies a renunciation of strict dynamic principles and the use of mathematical methods belonging to the theory of probability" (Rey 2017, p. 233, author's translation).

9 "[...] perfect randomness is, in its own way, a hypothesis as strong as that of absolute determinism and allows a very efficient mathematization. A rigorously non-deterministic world is perfectly probabilistic" (Rey 2017, p. 243, author's translation).

1.3.2. *Back to social sciences*

Armed with these too brief remarks on the elaboration of the concept of statistics in the natural sciences, it is possible to look back at the use of the concept of statistics in the social sciences. Indeed, it becomes possible to think that what society is – the object of the social sciences – remains in a *radical indetermination* as to the nature of its object, that the social sciences knew this from the outset in an implicit way since it is in them that the concept of statistics was born, but that they now know it *explicitly* because of the detour through the natural sciences.

Thus, for example, Durkheim, although a master in the use of statistics, eventually recognized that this use could not overcome the notion of society because society could not be broken down according to the gas kinetics model: he came to adopt an attitude quite similar to that of Exner and to consider that the *surplus* present in the notion of society was not to be sought on the side of a totality of society that would be greater than the sum of its members but rather on the side of principles making a *local* intelligibility possible, against a background of radical indeterminacy.

1.4. Conclusion

It has been shown above that just as Cassirer could both be part of the Kantian tradition and at the same time detach himself from it in order to better adopt the historical movement of knowledge, today there is a certain relevance in modifying the theoretical framework proposed by Cassirer in order to take into account the transformations of knowledge that he had not studied or had studied very little. The above discussion should therefore be considered as an attempt to lean on the spirit of the transcendental tradition while going beyond its letter.

1.5. References

Cassirer, E. (1906). *Das Erkenntnisproblem in der Philosophie und Wissenschaft der neueren Zeit*. Bruno Cassirer, Berlin.

Cassirer, E. (1923). *Einstein's Theory of Relativity*. The Open Court Publishing Company, Chicago.

Cassirer, E. (1938). Le concept de groupe et la perception. *Journal de psychologie normale et pathologique*, XXXV, 368–414.

Cassirer, E. (1939). *Axel Hägerström: Eine Studie zur schwedischen Philosophie der Gegenwart*. Wettergren & Kerbers, Gothenburg.

Cassirer, E. (1955). *The Philosophy of Symbolic Forms: Volume 1: Language*. Yale University Press, New Haven and London.

Cassirer, E. (1998). *Gesammelte Werke Hamburger Ausgabe (ECW)*. Birgit Recki, Hamburg.

Cassirer, E. (2013). *The Warburg Years (1919–1933): Essays on Language, Art, Myth, and Technology*. Translated by Lofts, S.G. Yale University Press, New Haven and London.

Hacking, I. (1990). *The Taming of Chance*. Cambridge University Press, Cambridge.

Kant, E. (2015). *Critique of Practical Reason*. Translated by Gegor, M. Cambridge University Press, Cambridge.

Lequan, M. (2007). Y a-t-il une morale transcendantale chez Kant ? *Revue de Métaphysique et de Morale*, 53(1), 115–139.

Rey, O. (2017). *Quand le monde s'est fait nombre*. Stock, Paris.

2

Leroi-Gourhan and the Birth of the Symbolic Function

2.1. The image of man

André Leroi-Gourhan (1911–1986) was professor of prehistory at the Collège de France (1969–1982). "Prehistory" means what comes before history. Life on Earth began almost 4 billion years ago, and the evolutionary lineage that led to man has been separated from other primates for 7 million years: all this without man and before man. Our difficulty in thinking about such a space of time comes from the fact that it overflows our own human time, which has given meaning to what we understand by "origin" or "genealogy". Thus, we not only make a genealogy from ourselves, but also from our own concepts of "Man" or "human nature". Against such an approach, Leroi-Gourhan insists on the fact that prehistory confronts us with a very different time, namely a "geological time". "Geological time" is that which follows the formation of mountains or the formation of species. In this sense, prehistory deals with "what is doubly buried in the earth and in the past" (Leroi-Gourhan 1964, p. 10[1]). It is necessary to excavate, find, date and interpret, and here the singularity of the researcher is that he is at the same time an archaeologist, ethnologist and paleontologist.

Paleontology is the science of fossils. Dealing with the buried – the unknown – it is an empirical science that is not experimental but retrospective (like history). Fossils bear witness to the evolution of living beings and thus to their kinship. The question is not "why is there evolution", but more empirically "how this evolution took place". The method is comparative, both to be able to reconstruct the living beings of the past from very meagre fragments and to compare species with each

---

Chapter written by Arild UTAKER.

1 All quotes from Leroi-Gourhan's books are translations by the author.

other. These are living beings that no longer exist – a paleontology, not an ontology – and which may have existed under conditions (e.g. climatic) other than those that prevail today. The dinosaur world no longer exists. It disappeared with the disappearance of the dinosaurs. Without this disappearance, the world might well be different and perhaps without the human species. Also Leroi-Gourhan's work is closer to Cuvier's – the founder of paleontology – than to Darwin's work. Cuvier's comparative method allows us to think of a "structural morphology" of the living world that accentuates a distinction between organs and functions[2]. This distinction later became the master distinction of Leroi-Gourhan.

A past separated from ourselves also gives us an exteriority. In other words, a critical distance from our own conceptions of life and particularly from that species of which we are a part, which the Swedish botanist Linné called "*Homo sapiens*". While human paleontology – paleoanthropology – does not exploit this critical distance, it risks being nothing more than a projection of our current conceptions of man. Thus, *Le Geste et la parole* begins with a critique of the "image of man" that "fossil men" seriously question. It is a question of highlighting the "philosophical situation of fossil man" (Leroi-Gourhan 1964, p. 166). It is up to "fossil men" to overturn the dominant image of man. In the first place, this image includes man in relation to a "natural man" from whom he comes or who represents his foundation or nature. The image has two versions – one "cerebralist" and the other "naturalist". The cerebralist version involves a man who, in coming out of nature, invents himself: "Rousseau, in *Discourse on Inequality* (1755), gives one of the first drafts of a 'cerebralist' theory of human evolution. The natural man endowed with all his present attributes, starting from the initial material zero, gradually invents, by imitating the beasts and by reasoning, all that in the technical and social order leads him to the present world" (Leroi-Gourhan 1964, p. 19). In short, there is a human species that comes out of nature where we find the "good savage" or "natural man". This conception is falsified by the discoveries of human fossils belonging to different human species. Instead of a single human species – our own – we should speak of human species in the plural, some of which can be found on a lineage leading to "*Homo sapiens*". Thus, "*Homo*" as the genus (family) of "*Sapiens*" takes on a whole new meaning. Instead of looking for a "missing link" that can link us to the apes, human fossils link us to human species that lived from about 3 million years ago; "it will take the 'Australopithecian revolution' of the last twenty years to unblock the question of the missing link" (Leroi-Gourhan 1964, p. 24). "Human paleontology has only exorcised the ape ancestor in the last few years" (Leroi-Gourhan 1964, p. 27).

---

2 Cassirer (1945) underlined this aspect in Cuvier. He compared Cuvier's "comparative anatomy" with Bopp's comparative grammar. Foucault (1966, p. 275) underlined the same relationship.

Who were the Australopithecus – literally "southern apes"? Neither apes nor modern humans, they belonged according to Leroi-Gourhan to the human race (to the "*Homo*"). Most paleontologists do not agree with this judgment. The dispute concerns the question of criteria. What criterion should define the human race? Leroi-Gourhan was formal in his approach: the criterion is given by the morphology of the body. The Australopithecus had a small brain but it walked upright, and the "standing position" is the definition of our genus. Also, Leroi-Gourhan argued that it was misnamed; "an improper name that goes back to the still recent time when we saw advanced apes. Here they are considered Australanthropes" (Leroi-Gourhan 1964, p. 94)[3]. On the other hand, it must be said that in the perspective of "geological time", the notion of species can be confusing in the sense that there are only almost imperceptible successions. But the criterion chosen is not arbitrary, and Leroi-Gourhan argued that there is a simple reason that excludes the Australanthrope from humanity: his brain is too small, only a third of ours. "No fossil relatively close to us leaves this feeling of strangeness, almost of embarrassment or discordance, none gives the impression of an inhumanized man more than that of an ape who would humanize himself. This discomfort comes from the fact that Australanthropes are in reality less ape-faced men than brain-boxed men defying humanity. We were prepared to admit everything except that we started with our feet" (Leroi-Gourhan 1964, p. 97). Later Lucy's discovery in 1974 proved him right; Australanthropes were bipedal. This was not only important against the "cerebralist prejudice", for the adventure of the human race is indeed the adventure of its body, of the slow formation of its body.

It must be stressed that Leroi-Gourhan took leave of the notion of "natural man". In fact, it has always been the starting point for speculation about the origin of man. According to Nietzsche, man and his soul are the product of a suppression of instincts that belongs to the natural man. In Freud's work, we find the great myth of "Totem and taboo" according to which the natural man becomes the civilized man by repressing the murder of the "primal father". And according to Darwin, man is basically a "natural man" in the sense that his abilities or qualities are gradually formed in relation to other animals[4]. On the other hand, man is formed by the "top", by his mind or brain as in Lévi-Strauss's (1963) view: "the original impulse which compelled men to exchange words must be sought for in that split representation that pertains to the symbolic function" (p. 62). The unasked question is therefore

3 Jacques Monod in *Chance and Necessity* agrees with him: "The Australopithecus, which Leroi-Gourhan rightly prefers to call 'Australanthropes'" (Monod 1970, p. 168, author's translation).

4 See Darwin (1996, p. 359): "In the distant future I see open fields for far more important researches. Psychology will be based on a new foundation, that of the necessary acquirement of each mental power and capacity by gradation. Light will be thrown on the origin of man and his history".

how the human body became a human body. And instead of wondering how to understand the evolution of the human brain, Darwin wandered into psychological speculations about feelings in humans and animals. For his part, Lévi-Strauss presented us with a symbolic function without genealogy. This is why Leroi-Gourhan's work represents a major advance over the dominant images of man; anthropology must be transformed into what did not exist in Darwin's time, paleoanthropology.

2.2. The human body

When one theoretically studies a living body, the crux of the matter is easily forgotten. Anatomy and physiology largely occupy the scientific scene, and the fact that a body breathes, eats, moves, flees, etc., becomes secondary. Curiously, a body's vital functions seem to be subordinated to the organs of the body. To explain the organs of a body is to explain the functions of a body. But it is always the organ that is primordial, and the function is comparable to a predicate that is subordinate to the subject of the sentence. The originality of Leroi-Gourhan's work was to dissect the living differently. A body has an organic (anatomical) side and a functional side. The two are linked, but they are not reduced to each other. Locomotion, nutrition (and grasping) and relationship (perception) are the three major functions: "(They) are linked from the first manifestations of animal life, so closely that any change in one of the terms presupposes those of the other two. More precisely, the functional whole corresponds to a rigorously synergistic whole. Although Cuvier, from the beginning of paleontology, in formulating the law of correlations, made this synergistic character of animal organizations perfectly clear, the analysis of the organs prevailed over the study of the functions or the functional whole" (Leroi-Gourhan 1983, p. 71).

Underlining the difference between organ and function implies that the living is doubly articulated. This is why "function" does not directly mean the function of an organ. It is not a question of saying that the function of the eyes is to see or that the function of the legs is to walk as if it were somehow a property or faculty of an organ. The eyes are organs, seeing is a function, and the function is not given in the organ that corresponds to it. Legs are organs, walking is a function and function does not automatically follow organs. In order to walk, the child needs to see others who walk standing up ("the wolf child" walking on all four legs), and to see, light is also needed (the child who does not see light – his eyes are glued shut, for example – becomes blind). A function can also move from one organ to another, such as the grasping that moves from the mouth to the forelimbs. Methodically, the distinction allowed Leroi-Gourhan to approach evolution from two perspectives: there is a "functional paleontology" (Leroi-Gourhan 1964, p. 57) and a "body paleontology" (of organs). He thus succeeded in highlighting lineages of evolution without going through zoological species. In relation to grasping, he divided mammals into two classes: those that had mouths as grasping instruments and those that also (but not exclusively) used the forelimbs. The former were mainly "walkers", and the latter mainly "arboreal" (such as apes). Compared to these two series, man represents an original solution: "The essential feature lies in locomotion: man has a foot that can be imagined as arboreal in the very distant past, but which, as early as the Australopithecus, appears to be adapted to walking on the ground. In other words, the human foot has undergone evolution in the same direction as walking mammals, whereas his hand has undergone the maximum possible evolution in the grasping directions. This anatomical paradox reflects the complete separation between the forelimb and the organs of locomotion, the standing position during walking, the vertebral straightening, characteristics which are all originally human" (Leroi-Gourhan 1983, pp. 75–76). But there are of course relationships between "the bodily device" and the "functional device" (Leroi-Gourhan 1964, pp. 57–58).

In order to fully understand the evolution, it must be stressed that this relationship was not synchronized. The forelimbs did not transform into hands for a later grasping function. The living being that became bipedal freed the forelimbs from locomotion. Also bipedalism is an anatomical cause of change of the forelimbs. It is a mechanical causality without purpose and without function. In a way, the hands are new organs compared to the old "body device". An organ is created without a previously given purpose. This distinction (between causes and functions (finality)) accounts for the chance linked to evolution, so that something new happens that is not contained in a previous phase. The hand – a new organ – thus subsequently opened up a new world through the functions that it gradually assumed. Here, there was no "design" – intelligent or mechanical – given in advance. It is better to say that the hand is the greatest surprise produced by evolution. It is the liberation of the hand (from locomotion) that is at the origin of the human adventure.

The function and the effects of the function came later. The hand profoundly transformed man's physical and functional apparatus. On a functional level, grasping moved from the mouth to the hand that became the exclusive grasping organ. There followed a "liberation" of the head and mouth. It was no longer used for grasping and then became lighter; the mandible is less strong and the teeth are smaller. The mechanical link between the mandible and the spine weakened and the skull (slowly) released a larger volume for the brain: "The arch literally opened like a fan. The deployment of the head range was not uniform (Figure 42): the forehead was contained in its proportions by the facial mass of which it provided the foundation, and it was not until '*Homo sapiens*' that the prefrontal cortex developed... The standing position, as early as the Australopithecus, has as its corollary an increase in the surface area of the head range in the middle fronto-temporo-parietal region" (Leroi-Gourhan 1964, p. 110). The change in the body implied a change in the skull, which in turn led to an enlargement of the brain. Obviously, the brain cannot be the cause of its own enlargement, nor can this enlargement be explained in terms of a pressure to develop an intelligence necessary for the survival of the species. As with the hand, the evolutionary path must be distinguished from the functions that the brain takes on once the "evolutionary ceiling" has been reached. Indeed, for the survival of the species, brain enlargement is rather useless as it requires too much energy. Evolution has created a living being with a brain that becomes far too large for its size.

In a recent book, *Qu'est-ce que le vivant,* Alain Prochiantz points out that in relation to our weight (comparable to the great apes) "we are 900 cubic centimeters too much": "The brain is used to move and connects the sensory world with the motor world. I move or remain motionless because I see or hear, feel, desire... Because of this, there is a linearity in primates, between the size of the body and the size of the brain. However, this relationship between the chimpanzee, the gorilla or the orangutan is lost as soon as one passes through the hominids ('Homo habilis', 'Homo erectus', 'Homo neanderthalis', etc.). The extreme point of this anomaly is reached with Sapiens, which has a brain of 1,400 cubic centimeters when 500 would be more than enough, given its size, for the sensory-motor functions of a basic primate. In short, to put it in a nutshell, we have 900 cubic centimeters 'too much'". (Prochiantz 2012, pp. 85–86, author's translation)

Also, the brain played its major role only with the advent of "*Homo sapiens*". "It is surprising that the importance of the brain's volume only comes into play afterwards. In reality, it is difficult to give pre-eminence to this or that character, because everything is linked in the development of species, but it seems certain to me that brain development is in some way a secondary criterion. It plays, when mankind is acquired, a decisive role in the development of the species, but it is certainly, in terms of strict evolution, correlative of the standing position and not, as has long been believed, primordial" (Leroi-Gourhan 1964, p. 33).

The opposite position implies an amalgamation of organs and functions. An organ is explained by its function and a function by its organ. Thus, a functionalist explanation becomes common in biology. In Darwin's view, the function of an organ is explained by virtue of its usefulness in the face of environmental pressure. Explaining the evolution of the brain in the manner of Leroi-Gourhan is therefore excluded and given the immediate non-utility of this evolution, this enlargement would not have taken place. Why did human beings and their brains grow during their evolution? Is there a utility or an advantage that can explain it? Or is it due to mechanical causes, such as a change in food, the control of fire to prepare food, or climate change? Leroi-Gourhan's interest is in opting for the latter. It follows that the theory of "natural selection" cannot explain the evolution of the brain. It is not increased in function of a utility because at the beginning its increasing volume was rather an obstacle (requiring too much energy). In fact, Darwin, so preoccupied with utility, projected a current utility of an organ on what should explain the emergence of this organ. He thus assumed a continuity of evolution that enclosed it in a "metaphysics of utility" explaining everything and nothing.

The "biological originality of man" (Leroi-Gourhan 1983, p. 68) is an originality that is not visible from a physiological or anatomical perspective because the primordial fact is the morphology (shape) of the human body; its standing position, hands free while walking, a body that can run and throw stones (at the same time), etc. (Leroi-Gourhan 1983, p. 68). In fact, form (morphology) is the link between organs and functions. Thus, the morphology of the body is the context accounting for techniques, from biological fact (in the morphological sense) everything that concerns man derives; techniques, language, thought and images. Placing man in the evolution of living beings places him in the natural order but in a way that radically breaks with our usual conceptions of the human being.

2.3. The hand and the tool

In the course of human evolution, the hand became the grasping organ and, by its form, it became a tool itself; the human being does something with his hands. Technicality is not added to the hand, but is linked to what a hand can do: "The tool does not appear intrusively as if it were superimposed on an operational framework already virtually constituted, it emerges in a way from the hand in the very movement of its liberation" (Leroi-Gourhan 1983, p. 253). This is why the making of a tool is not the product of thought: "On a philosophical level, the problem of human technicality could thus be considerably modified; it would appear not as a consequence of 'intelligence' with its currents and waves, but as the result of the accession to a highly organized motricity, as the product of a new bodily conditioning. Technicity would not wait until the thinking brain has already accomplished a long ascent; it would only take advantage of the evolution of increasingly richly organized brain territory to take on more and more reflective forms" (Leroi-Gourhan 1983, p. 250). This is not to say that the tool is exclusively human (of course not), but to understand that for 2 million years the development of tools followed a "geological rhythm". From the primitive flint to the two-sided flint, the rhythm was the same as that which followed the enlargement of the brain's volume: "Up to the Upper Paleolithic threshold, the course of progress seemed to be molded on the 'geological' rhythm, and no longer on the 'historical' rhythm. It is quite obvious that there was no discontinuity between the biological and the technological, but rather an increasingly accelerated divergence" (Leroi-Gourhan 1983, p. 255).

It goes without saying that this development was not just an effect on the hand. More profoundly "the tool appears as a true anatomical consequence, the only outcome of a human being who has become, in terms of his hand and his teeth, completely inert and whose brain is organized for manual operations of a complex nature" (Leroi-Gourhan 1964, p. 129). The tool is linked to its survival because with the tool the hand could obtain a strength and technique that it did not possess alone. It extended the hand but not in an artificial way since it is part of the hand and its technique. "A stone is shattered by a hand and the hand thus becomes human" (Leroi-Gourhan 1983, p. 70).

On the other hand, the hand became a human hand by the fact that the tool became detached or dissociated from it. The hand leaves the tool in its place and it retains its character as a tool; it is not just any stone. The tool that extends the hand is also a tool that detaches from the hand: "The human hand is human by what detaches from it and not by what it is: a fairly simple osteo-muscular device..." (Leroi-Gourhan 1965, p. 41). Detached from the hand, the tool is removable. If it extends the body, its character as a tool also puts it outside the body. In the animal, the tool and the organ merge. "The crab's claw and its mandibular parts merge with the operating program through which the animal's acquisition behavior is expressed" (Leroi-Gourhan 1965, p. 35). In humans, on the other hand, the removable tool is an external organ. In this sense, the tool is the first element of the human functional system that is externalized. After the liberation of the hand (the liberation of an organ) came the "liberation" of the tool, which was a technical "liberation" within the functional system of man: "All human evolution contributed to placing outside man that which in the rest of the animal world responded to specific adaptation. The most striking material fact was certainly the 'liberation' of the tool..." (Leroi-Gourhan 1965, p. 34). The fact that the tool made by the hand is removable, that is to say, in an exteriority with respect to the human body, implies that technical gestures can be repeated with the same tool, which will thus be linked to a memory and a project (for example, the project to use the same object later). The tool separates both from the body and from an external object to which it is applied. More fundamentally, before the tool – as a removable object – there was no such separation between the body and the world of objects.

This is why the externalization of the tool does not refer to an interiority that pre-existed it. It is primordial. However, exteriority is only one aspect of the mode of being of a technical object. It fully exists only in a technical fact and a technical gesture. The tool is only a tool inside a gesture that manipulates an object in a temporal sequence where the before and after (one must do X before Y, etc.) are crucial: "Technique is both gesture and tool, organized in a chain by a real syntax that gives the operating series both their fixity and their flexibility" (Leroi-Gourhan 1964, p. 164). Also, the technical gesture creates a "technical intelligence" linked to a memory. Although it is instrumental, it is a mistake to be blinded by the instrumentality of a technical gesture. What is crucial is what a technical course means and what the category of tool as well as that of technical gesture imply in general for the human being. In fact, it is the technical gesture that in the evolution of Man will transform his functional device. It is the technical gesture that opens up to a non-genetic memory and a verbal language.

The technical gesture linked to a flexible syntax is rhythmic: "One of the operating characteristics of humanity, from its earliest stages, has been the application of rhythmic percussion, repeated at length. This operation is even the only one that marks the entrance of Australanthropic humanity, since it left as traces the choppers of shattered pebbles and polyhedral balls born from long hammering. From the outset, the manufacturing techniques are set in a rhythmic atmosphere, at once muscular, auditory and visual, born of the repetition of the gestures of shock" (Leroi-Gourhan 1965, p. 135). Rhythmic, the technical gesture is something other than a reaction of the hand in the face of danger. In terms of grasping, it assumes a certain independence from the other two functions; locomotion and relationship (perception). As the tool was externalized, the gesture became autonomous. This development can also be explained from the anatomy of the human body: "Throughout his evolution, from reptiles, Man appeared as the successor of those between creatures that have escaped specialization. Neither his teeth, nor his hands, nor his foot, nor finally his brain, reached the high degree of perfection of a mammoth's teeth, a horse's hand and foot, the brain of certain birds, so that he has remained capable of almost all possible actions, that he can eat practically anything, run, climb, and use the improbably archaic organ in his skeleton, the hand, for operations by a brain overspecialized in generalization" (Leroi-Gourhan 1964, p. 168). The hand that surpasses itself in technical gestures more generally shows a relative dissociation between the three moments of the functional device of mammals. It is no longer primarily a perception that triggers locomotion or grasping. The functions assume, on the other hand, a certain independence with respect to the organs with which they are associated in order to enter into synergy with the other functions. It is in this sense that the technical gesture appeals to the face – to the eyes and the voice.

2.4. Technique and language

The morphology of the human body implies that the hand–face relationship plays a fundamental role. From the technical gesture, the functions related to the face are transformed. It calls upon a gaze that knows how to follow it and a voice that knows how to memorize it. This presupposes a head freed from grasping and more precisely a mouth freed for speech. Leroi-Gourhan's argument is that the development of techniques was impossible without a certain language. The problem is that speech has left no fossils. How can we carry out a "paleontology of language" without a call to a mysterious origin? Instead of seeking an origin, he undertook "the account of the geological relations between technique and language" (Leroi-Gourhan 1964, p. 11). Without "fossil languages", a comparison with apes was necessary in order to better define the hand–face relationship, essential for a "paleontology of language". "The ape works with its lips, teeth, tongue and hands, just as present-day man speaks with his lips, teeth, tongue and gesticulates or writes with his hands... In other words, from a formula identical to those of the primates, man makes concrete tools and symbols, both of which are part of the same process or rather use the same basic equipment in the brain" (Leroi-Gourhan 1964). The plasticity of the brain – which was created in the course of its evolution – has given rise, from the hand-face relationship, to the cerebral link between tool and language. They are "linked neurologically" (Leroi-Gourhan 1964, p. 163), just as, in terms of thought, "the hand and the voice remain closely united" (Leroi-Gourhan 1964, p. 298).

Hence, language has developed in line with the evolution of techniques. "At those stages when the comparative study of tools and skulls seems to show that the industry developed at a rate corresponding to that of biological evolution, the level of language could only be very low, but it certainly exceeded the level of vocal signals" (Leroi-Gourhan 1964, p. 163). While the gorilla possesses vocal signals, it is the character of the removable and externalized tool that seems to imply a human speech that breaks with the world of signals. "Indeed, what characterizes 'language' and 'technique' in apes is their spontaneous appearance under the effect of an external stimulus and their no less spontaneous abandonment or failure to appear if the material situation that triggers them ceases or does not manifest itself. The manufacture of the chopper or the biface is a very different mechanism, since the manufacturing operations pre-existed at the time of use and since the tool persisted with a view to subsequent actions" (Leroi-Gourhan 1964, p. 164). It is precisely this specificity of the technique that imposes a 'technical intelligence' that through a word can cover 'the operating chains' and also recognize the tools that can be used later. "These are technical activities that are unthinkable without a verbalized intellectual fixation" (Leroi-Gourhan 1964, p. 298).

Even if the tool can be seen as the first symbol just as technique is already a kind of language (or embedded in a language), the essential thing is that the language developed from the technique in a way that demonstrates a hijacking of it in the end. The technical gesture and the tool prefigured the very form of speech as an orality that is distinct from a communication of signals. As the technical gesture implied an externalization of the hand through the tool, speech implied an externalization of the mouth through the verbal symbol: "Speech is a verbal tool, isolable from the mouth that emits it, just as the manual tool is isolable from the hand" (Leroi-Gourhan 1983, p. 81). As the tool came from the hand, speech came from the mouth. A cry is part of a mouth that can signal danger but a word is only a word by being detached from the speaking voice. Without such an exteriority, it is not shared and heard as language. Like the tool, language is "removable" and externalized.

To understand this, the mouth in insolation can confuse us. In fact, speech is isolated from the mouth by means of the ear. Speech does not occur if it is not initially related to the ear or hearing. The sound emitted is sound only received or heard. That is why speech also does not refer to an interiority that pre-exists it and that it is supposed to express. But where the tool works an object, speech works nothing – except itself. There is not an organ of speech and the birth of language was not from an organ, either the vocal apparatus or the brain. If we speak of an "organ of language" (as in Chomsky's case), we do not admit the essential distinction between organ and function. For Leroi-Gurhan, there is a linguistic and symbolic function, not linguistic and symbolic organs. Such a function is of course articulated on the body. The relationship between articulation (mouth) and perception (hearing) has a physiological basis in the relationship between voice and hearing. Speech is articulated on this relationship and becomes possible when perception dominates (and governs) the articulation. This phenomenon can be called self-affection in the sense that with speech the body affects itself.

The exteriority of speech also distinguishes it from the world (otherwise it would have been only a signal). The tool and the word imply a distance "between man and the inner and outer environment in which he is immersed". This detachment "is expressed in the separation of the tool from the hand and the word from the object" (Leroi-Gourhan 1965, p. 33). "The most striking material fact is certainly the 'liberation' of the tool, but in reality the fundamental fact is the liberation of the verb and this property of placing its memory outside itself, in the social organism" (Leroi-Gourhan 1965, p. 34). In short, a gap appears between man and his environment, which characterizes "the faculty of symbolization, or more generally this property of the human brain, which is to maintain a distance between the lived experience and the organism that supports it" (Leroi-Gourhan 1965, p. 33). It should be emphasized that this distance or gap (from the world) is not simply postulated. On the one hand, it is an externalization of the body that produces a distance both from the body and from the world and, on the other hand, what is externalized – the tool, the word – is precisely what affects the body. Instead of a relationship that goes from the body (the subject) to the world or from the world to the body, the characteristic of a human world consists of the circularity that characterizes its functional device: externalizations that in turn can only affect and shape the human body. By the same movement, it is also the world that becomes an external world through the tool and through speech: "In other words, since at a human scale the technical function is externalized in the removable tool, since the perceived object also becomes external in a verbal symbol, the movement in all its visual, auditory and motor forms is also liberated and enters into the same evolutionary cycle" (Leroi-Gourhan 1965, p. 87).

2.5. Language and visualization

After the tool and the word, a third step led directly to *Sapiens*. Next to flints or tools, the paleontologist finds new types of objects: ornamental objects, dyes (such as ochre) or objects with lines carved into the bone or stone. It is as if the word goes back to the hand and transforms it into a graphic hand, i.e. a hand that draws and traces. Let us quote: "Graphic figuration was born with the first development of '*Homo sapiens*', which is a precious indication" (Leroi-Gourhan 1965, p. 209). For "nothing comparable to the tracing and reading of symbols existed until the dawn of '*Homo sapiens*'" (Leroi-Gourhan 1964, p. 262). In the beginning, this figuration (visualization) was directly related to speech or, more generally, to orality. This is why it is strictly speaking non-figurative and why it is rather about "the oldest rhythmic manifestations expressed; no precise meaning is any longer graspable in these very modest witnesses" (Leroi-Gourhan 1964, p. 263). Also, "the oldest known figures do not represent hunting, dying animals or touching family scenes; they are graphic pegs, the material of an oral context that has been irretrievably lost" (Leroi-Gourhan 1964, p. 266).

One could say that they are figures that serve to give a foundation to speech, a bit like the rhythmic series of sticks. In any case, it is a form of language and not ornaments without deeper meaning. As a language, figuration also brings into play the hand–face couple in the sense that the graphic hand refers to vision: "On the two ends of the operating field, two languages are constituted, from the same sources, that of hearing which is linked to the evolution of the territories coordinating sounds, and that of vision which is linked to the evolution of the territories coordinating gestures translated into graphic symbols" (Leroi-Gourhan 1964, p. 270). "It is therefore obvious that from the source, phonation and graphics serve the same purpose" (Leroi-Gourhan 1964, p. 269). The phonic gesture and the graphic gesture express (both) something. While one is related to hearing, the other is related to seeing. This means that to the ear that hears (understands) corresponds eyes that read. The phonation-graphics couple thus refers to the same transformation of human vision and hearing. It is important to underline this since it is easy to forget that figuration (graphics in general) refers to a vision that can read, not in the sense that one reads a piece of writing (this happens much later), but in the sense that the vision reads and understands traces or marks – paintings. It follows from this that neither the word nor the figure are "layers of reality". The image or figure exists only through the graphic hand, just as the word exists only in the articulation–hearing relationship. This is why the first figurative art "was a symbolic transposition and not a layer of reality, that is to say that there was between the line in which one admitted seeing a buffalo and the buffalo itself the distance that existed between the word and the object" (Leroi-Gourhan 1964, p. 266). The expression "and not a layer of reality" means that the figure is an externalization like that of the tool and the word. The technical gesture, the verbal gesture and the graphic gesture are the three externalizations that mark the functional device of man.

How can we explain a vision that can read without reacting immediately to what it sees or to what moves (a danger or a promise of food)? First of all, with a raised head turned toward the horizon and not toward the ground, vision and hearing are dominant senses in humans. "In man, among the senses of relationship, olfaction is in a special position. Indeed, vision and hearing engaged in language, like the hand, enter alone into the system of transmission and reception which makes the exchange of figurative symbols possible. Olfaction, purely receptive, has no additional organ for the emission of odor symbols. It remains foreign to the most characteristic human device" (Leroi-Gourhan 1965, p. 112). A dog's world is perhaps one of smells. Olfaction does not refer to an organ that knows how to express a smell, whereas vision refers to an organ that knows how to express what we see, namely the graphic hand or the verbal gesture. Thus, a reception (perception) translates into an emission. But an emission (through speech, through the graphic gesture) only exists when it is perceived (received) and this means that the "system of emission and reception" refers to a more fundamental fact linked to the articulation–perception couple. On the other hand, this shows the originality of olfaction as is a perception unrelated to an articulation in a sense that can open an "involuntary memory" (as stressed by Marcel Proust).

The articulation of the technical gesture is linked to its perception through vision, and the articulation of speech is linked to its hearing. Here, there is not a motricity that conditions what is done, which is the cause of the result of the motricity. In general, it is the sensory–motor schema (linked to causality) that is excluded: "If in the technique and language of the totality of Anthropians (the human species before Sapiens) motricity conditioned the expression (as, for example, the primitive tool), in the figurative language of the most recent Anthropians reflection determined the graphics" (Leroi-Gourhan 1964, p. 262). The motricity of the hand is not enough, it is transformed into a graphic hand by its link to reflection. Here "reflection" means both perception and reflection, as if what the hand produced was reflected in the vision or a "reflective perception". Therefore, perception (reflection) dominates the articulation of the graphic hand, just as hearing dominates the speaking gesture. This relationship is not a cause–effect relation. Rather, there is a simultaneity between the graphic hand and the perception that follows it; the gaze dominates the drawing hand.

Thus, vision and hearing open up to a "symbolic reflection" that the other senses do not recognize (Leroi-Gourhan 1965, p. 209). Images and words are the "symbolic reflections" of hearing and seeing. It is through the specificity of seeing and hearing that man reflects reality in verbal, gestural or figured symbols (Leroi-Gourhan 1965, p. 206). At the same time, he can only reflect himself in his own externalizations, that is, seeing, hearing and feeling himself. The image or the word is also a mirror. Leroi-Gourhan notes in this connection the fascination with objects that shine and reflect; "the crystals that cast fires directly touch the depths of man's reflected thought" (Leroi-Gourhan 1965, p. 214). For "the unusual in form, which is a powerful source of figurative interest, exists only where the subject confronts an organized image of his universe of relation with the objects that enter his field of perception" (Leroi-Gourhan 1965, p. 214). The image and the figurative also represent a difference in relation to speech:

> Graphic symbolism benefits, in relation to phonetic language, from a certain independence. Its content is expressed in the three dimensions of space what phonetic language expresses in the single dimension of time. At the level where we are still situated, the link between language and graphic expression is one of coordination and not of subordination. The image then possesses a dimensional freedom that will always be lacking in writing (Leroi-Gourhan 1965, p. 272).

The image is "radiant" and there is a profound link between multidimensional graphics and mythology. In general, mythology, ritual and magic are not distinguished from language but are part of it. It is we who make the distinction, since our way of thinking is very different from that of the Paleolithic. Leroi-Gourhan observes: "The longest part of the evolution of '*Homo sapiens*' took place in forms of thought that have become foreign to us, even though they still underlie an important part of our behavior. While we live in the practice of a single language, whose sounds are inscribed in a writing associated with them, we can hardly conceive of the possibility of a mode of expression in which thought graphically has a sort of radiant organization" (Leroi-Gourhan 1965, p. 273). Mythology and esthetics here refer to each other in a way that expresses emotions and emotional bonds: "The fundamental link between art and religion is emotional, but not in a vague way, it is closely linked to the conquest of a mode of expression that restores the true situation of man in a cosmos in which he is inscribed as the center and which he does not try to penetrate by the line of reasoning in which the letters make thought a penetrating line, of long range, but as thin as a thread" (Leroi-Gourhan 1964, pp. 273–275). It is the "penetrating line" of the writing of characters that presupposes our logic and that prelogical thought does not know[5].

5 If our logic is linked to writing, Leroi-Gourhan (1964) distinguishes it from primitive graphics or symbolism: "On the other hand, the origin of writing has often been linked to the processes of memorizing numerical values (regular notches, knotted strings, etc.); if, indeed, alphabetical linearization may have, from its origin, maintained relations with numbering devices that were necessarily linear, it is not the same for the oldest figurative symbolism. This is why I am inclined to consider pictography as something other than a childhood form of writing" (p. 270). This is why Jacques Derrida is mistaken about Leroi-Gourhan when he refers to him to explain his own notion of "generalized writing" (Derrida 1967, p. 125). In contrast to Leroi-Gourhan, Derrida remains "logocentric".

In other words, paleontology exempts us from starting with thought and logic. Here everything begins with the body and the senses of the body, more precisely with a vertical and bipedal body. Otherwise, we link vision or language to thought (to the intelligible), forgetting the hand and the technical gesture. Technique – the "techné" – thus becomes subordinate to thought, and our philosophy ends up oscillating between mentalism and cerebralism. It should be added that this system of thought has no place for a "functional device" that is inserted between man and his world. The same is true of the neo-Darwinian paradigm where functions are swallowed up by their organs – just as the brain swallows up thought. Perhaps our difficulty here is linked to our dominant conception of language that dissociates language from perception and articulation.

In short, the human body becomes a human body by freeing itself. In this way, its functional device is created, articulated on its organs but without being reducible to the organs of the body. It is rather a rhythmic body: "The functional whole of man is also a rhythmic device" (Leroi-Gourhan 1964, p. 263), thus a body that moves and dances. This device can be considered as a triangle (Leroi-Gourhan 1965, p. 86) that is made up of three extremes: technical gesture, verbal gesture and graphic gesture; the tool, the word and the figure. But given the intrinsic relationships between the three extremes, it is more accurate to say that this device can be interpreted in three different ways: from the symbolic, gesticulation, speech and figures are part of language. From the technical, it refers to figures and speech as its necessary dimensions. And from the esthetic point of view, figuration and rhythm dominate.

This device is of course not antinomic in relation to the body. It should be recalled that the body is both organic and functional; doubly articulated in a way that characterizes the living in general. When the functional device can be characterized as grammatical (Leroi-Gourhan speaks of syntax), the anatomical body follows the mechanics of cause and effect. In this context, the brain is not an organ like any other. It is the organ that connects and coordinates functions; an organ of coordination and interaction. It "doubles" speech – as one of its materials – but it is not the cause of speech. Direct causality only intervenes in the case of a dysfunction. Thus, we can perhaps propose the hypothesis that due to the brain (and its plasticity) extended by a device that is both distinct and linked to it, man is first and foremost "Grammatical Man" and that there is no "Neural Man" who can explain this.

2.6. Memory and history

The problem of human evolution is simply memory. In order for human groups to survive it is not only necessary that they develop techniques, but that these techniques are transmitted. In short, the challenge was to create a collective memory that both provides a link to the past and consolidates the group through a common memory. From "*Homo sapiens*", the constitution of a social memory apparatus dominates all the problems of human evolution (Leroi-Gourhan 1965, p. 24). It is evolution that required man – alongside his genetic memory – to create a new and unique memory, a collective memory. This memory no longer belongs to the species but to a group: "Breaking the link between the species and memory appears to be the only solution (and a human-only solution) that leads to rapid and continuous evolution" (Leroi-Gourhan 1965, p. 23). This is why *Sapiens* mark the advent of a historical time different from "geological time". Speaking of societies without history or before history rather shows our way of linking history to change. But the first societies up to the Neolithic revolution – agriculture – had found the means to maintain and transmit their memory: we must repeat ourselves. That for 30,000 years not much happened shows the success of the first *Sapiens* groups.

The main instrument of non-genetic memory is language. It is perhaps at its core a memory technique – a mnemonic. In any case, it is a condition of our memory. On the one hand, language is in itself transmission, it only exists as a transmission of the past, and, on the other hand, through speech, myths maintain the link to the past and to the knowledge that is transmitted from ancestors: "once upon a time". Thus, the past assumes a reality that without language could not exist. Language comes to us from the past and with language we speak of the past. In short, language is the historical phenomenon *par excellence*. This of course presupposes the exteriority of language, for the memory of which Leroi-Gourhan speaks is an externalized memory; it belongs to the group. After the "liberation of the hand", we have the "liberation of memory". If the genetic memory proper to the species is given in the individuals that constitute it, the memory proper to man, on the other hand, is given (externalized) in his group. His memory is placed outside himself, in the social organism or in its functional system. The tool, the word and the figure are at the same time bearers of a memory and produced due to the same memory. When they change, the shape of the memory changes as well.

History implies that ethnic groups replaces species (Leroi-Gourhan 1965, p. 12): hence the dispersal and differentiation of *Sapiens* in different corners of the globe. Similarly, technological evolution becomes more important than biological evolution. Starting from the founding triangle (above), history starts to become "dynamic" from the technical externalizations of tools, speech and figures (the images). Tools are externalized in machines, speech is externalized in writing or letters, and images are externalized by visual technologies. Later these externalizations will find a common language through electronics, and they are somehow coming together on the computer screen. At this stage, it is perhaps the brain itself that will externalize itself in a supercomputer; the memory being transformed into mechanical memory like the memory of a computer. Thus, we have three successive externalizations of memory through orality, writing and finally through electronics.

2.7. Conclusion

In this situation – ours – the body as it was as the *Sapiens*' starting point risks becoming obsolete. Subsequent externalizations (liberations) risk becoming autonomous in a way that no longer depends on a body. Rather, the body becomes a somewhat embarrassing supplement to an external brain. So what is the fate of "*Homo sapiens*"? "Freed from his tools, his gestures, his muscles, the programming of his acts, his memory, freed from his imagination by televised means ... the '*Homo sapiens*' of zoology is probably near the end of his career ... . The great problem of the world already present is to be solved: how will this obsolete mammal, with the archaic needs that drove its ascent, continue to push its rock up the slope if all it has left one day is the image of its reality" (Leroi-Gourhan 1965, p. 266). Leroi-Gourhan's scenario is not so much that man has become a well-programmed machine. Rather, the hypothesis is that man has externalized himself into machines guided by the same forms of programs. In this sense, his body no longer has a real or perceived reality, but exists only as an "image of his own reality" – thus unrealized.

If one has difficulty understanding this possibility, there is of course a reason. Perhaps our dominant ways of thinking fits perfectly with such a possibility. We need only mention the notion of program. Programs produce images, texts and goods. Not to mention genetic programs. An American philosopher, Daniel Dennet, has even conceptualized natural selection according to Darwin as a program that explains evolution in the form of algorithms (Dennet 1995, p. 82). Against such a theory, empirical research – with empirical results – that show that evolution is much more complicated seems to be falling flat. In conclusion, the neo-Darwinian paradigm is weak empirically and strong theoretically by assuming a dominant system of thought making the following question obligatory: "Which genetic mutations have led to the increase in brain size?" (Prochiantz 2012, p. 102, author's translation). Never the less Prochiantz adds: "Let us note here that this cerebral increase may not be independent of other changes... I could have written about bipedalism and the liberation of the hand" (Prochiantz 2012, p. 102, author's translation). But even if he is unable to do that, his great merit is to try – to try to crack the neo-Darwinian framework. As the following quotation shows: "The physicalist and mechanistic ideologies that have, until recently, imposed a straitjacket on the life sciences" (Prochiantz 2012, p. 140, author's translation). He tries but he can't – this system of thought is too strong even for him. To achieve this, we must celebrate the divorce between molecular biology and paleontology (the biology of evolution). Stephen Gould has moved cautiously in this direction, but for Leroi-Gourhan this marriage did not even take place. That is why he occupied an invisible theoretical field.

As we have underlined, he excludes a utility that can explain the evolution of the brain or language. Assuming such a utility, there would be no problem of meaning or sense. One becomes incapable of explaining the origin of language except by the myth that poses genetic mutations. Alternatively, we can affirm that this origin is indeed a "mystery". This incapacity shows the other side of a conception of language that excludes the symbolic in order to reduce language to the digital or to a "syntactic machine" in the brain. The danger that Leroi-Gourhan points out about the externalizations of our brain and of our language thus becomes invisible to a mainstream of modern thought that thinks about language on the basis of such externalizations. It follows that a specificity of man does not exist. On the one hand, we only extend what is found in animals; on the other hand, we are extended by new technologies. So the symbolic does not exist. Man can disappear and this disappearance can be supported (or encouraged) by our ignorance and blindness. This is why Leroi-Gourhan is more topical than ever. He reminds us that the symbolic function conceived as "the ability to reflect reality in verbal or gestural symbols, or materialized by figures" (Leroi-Gourhan 1965, p. 214) is proper to man.

2.8. References

Cassirer, E.A. (1945). Structuralism in modern linguistics [Online]. *WORD*, 1(2), 99–120. Available at: https://doi.org/10.1080/00437956.1945.1165 9249 [Accessed 19 August 2019].

Darwin, C. (1996). *On the Origin of Species*. Oxford University Press, Oxford.

Dennet, D. (1995). *Darwin's Dangerous Ideas*. Penguin Press, London.

Derrida, J. (1967). *De la grammatologie*. Éditions de Minuit, Paris.

Foucault, M. (1966). *Les mots et les choses*. Gallimard, Paris.

Leroi-Gourhan, A. (1964). *Le Geste et la parole. Vol. 1 Technique et Langage*. Albin Michel, Paris.

Leroi-Gourhan, A (1965). *Le Geste et la parole. Vol. 2 La Mémoire et les Rythmes*. Albin Michel, Paris.

Leroi-Gourhan, A. (1983). *Le fil du temps*. Fayard, Paris.

Lévi-Strauss, C. (2008). *Structural Anthropology*. Hachette, London.

Monod, J. (1970). *Le hasard et la nécessité*. Le Seuil, Paris.

Prochiantz, A. (2012). *Qu'est-ce que le vivant?* Le Seuil, Paris.

3

Simondon, Language and Technology

The philosopher Gilbert Simondon (1924–1989) gained a solid reputation as a specialist in technology with the publication of *On the Mode of Existence of Technical Objects* (Simondon 2017), his complementary thesis, in French in 1958. In addition, he devoted numerous studies to technology, most of which are collected in (Simondon 2014). On the other hand, he did not write any in-depth analysis of language. Our colleague and friend Ludovic Duhem has endeavored to shed light on the reasons for this absence (Duhem 2015). By noting, in a vast corpus, all of Simondon's remarks and analyses on language, he arrived at two conclusions that we take up again (by reformulating them):

– the absence of a specific analysis does not reflect Simondon's ignorance or indifference to the topic, but arises from the secondary status of language in his philosophy of nature. For Simondon, language is *not* the original condition of meaning. To say this in a language that is not his own, Simondon departed from logocentrism (defining the nature of the human being through language) and criticized the scholastic attitude of philosophers who focus their attention on language because it is the only field of action within their reach;

– while the question of language is nevertheless encountered by Simondon, it is at the end of a general survey on communication processes (Simondon 2010). This survey begins with the study of physical communication phenomena, and continues with the study of ethological data on communication modes within and between species, before leading to the study of human communication. The latter cannot be reduced to articulated language: Simondon insisted on the need to take into account the multiple channels and technical functioning of communication, of which articulated language is only one particular mode and not the paradigm.

Chapter written by Vincent BONTEMS.

In addition to these two conclusions, which already provide a better understanding of the reasons for the asymmetry of processing and the imbalance in the work, we will add a third consideration: Simondon believed that technology had, in terms of law and fact, a universality superior to that of any natural language (Simondon 2014, pp. 269–278). It is the technical object that is the material and symbol of transindividuality, because it universally conveys information more than any speech or text, which always implies a particular coding. If texts (such as the inscriptions in linear A) may perhaps remain a dead letter forever, it is because we do not have the means to decipher them and do not have the elements, suggesting a translation into a language we already know; it would, on the other hand, be absurd to postulate that the functioning of an unknown technical object can indefinitely escape reverse engineering. The universality of technology transcends cultural differences. Simondon was aware that, for a long time, technology only made sense through its ritual insertion into a particular culture (Simondon 2014, pp. 315–330), but he considered industrialized technology to be more "ecumenical" than any other cultural production.

This makes it all the more interesting to determine under what conditions language can give reason to technology, its organization, its evolution and its values. To offer a first point of reference on this subject (which deserves a much more in-depth study), we will consider three perspectives on the articulation of technology and language in Simondon's work:

1) first, to provide some elements on the priority given to the investigation of technical operations and structures over those of language;

2) then, examine how Simondon develops a "technological" vocabulary on his own behalf in order to characterize technical objects and their filiations;

3) finally, we question the inadequacy of this language to guide the investigation of "technical lineages" and propose the use of diagrams.

3.1. The precedence of technology over language

Before specifying what kind of precedence Simondon gave to the question of technology, in relation to that of language, it should be recalled that, more generally, he defined philosophy as an investigation into the nature of beings prior to that of the possibilities of language:

> [...] it is absolutely insufficient to say that it is language that allows man to access meanings; if there were no meanings to support language, there would be no language; it is not language that produces meaning; it is only that which conveys information between subjects that, in order to become meaningful, needs to meet that *apeiron*

> associated with the individuality defined in the subject; language is an instrument of expression, a vehicle of information, but not a creator of meanings. Meaning is a relationship of beings, not a pure expression. (Simondon 2005a, p. 307, author's translation)

We will not attempt to justify this position here but to clarify its implications: if meanings pre-exist language, it is therefore the "being of meanings" that constitutes the first object of the philosophy of nature, and not the "meanings of being" that come under a philosophy of language. An exchange between Simondon and Paul Ricœur at the meeting of the French Society of Philosophy on February 26, 1960 testifies to his radical opposition to the perspective of the philosophy of language (of which hermeneutics is only a variant):

> Mr. Ricœur: I don't see how we can constitute the world of discourse starting from the natural region, which is itself something in the discourse.
>
> Mr. Simondon: No. If we admit that the natural region is part of the discourse, we can't. But there is an assumption. [...] But how can we admit that nature is a part of the discourse? That is the postulate that precedes your argument, and that I will absolutely reject. (Simondon 1960, p. 759, author's translation)

Philosophers of language certainly claim their modesty: they limit themselves to clarifying linguistic meanings, but since they posit that all meaning finds its condition of possibility in language, they are, in fact, asserting the lofty privilege of subjecting all meanings, whatever their origins, to the demands of categories that they base solely on the more or less rigorous study of language structures. By rejecting this claim, which is, so to speak, logo-morphic in the name of a "naturalist" requirement (Carrozzini 2015), Simondon also rejected the ontology of traditional philosophy, often nothing more than a kind of sublimation of grammatical or logical structures – a point that Ricœur himself had identified in connection with Plato and Aristotle (Ricœur 1960) but to which he found nothing to reproach.

"Technology"[1] cannot therefore adopt a linguistic paradigm. Taking a stand in the heated debate[2] in 1955 between Claude Lévi-Strauss, a proponent of linguistically inspired structuralism, and André-Georges Haudricourt, who defended the need to study artifacts independently of language in order to understand a

1 For a study of Simondon's conception of technology, see Bontems (2015a, pp. 7–27).

2 The terms of this debate are summarized in Jean-Yves Château's "presentation" (Simondon 2005b, pp. 52–55).

people's culture, Simondon described the agenda of "comparative technology" in these terms:

> The technical object can provide conceptual models other than and independent of linguistics. The systematic vision of the human world would probably be modified, because structuralist formalism generalizes a classifying and categorical thinking that is only one aspect of inter-human relations; on the contrary, technology reveals a *relational* function of the coupling between organism and environment, man and world, developing dialectically from level to level, and thus giving meaning to the norm (or pattern) of progress, according to a realistic epistemology, different from structuralist nominalism. (Simondon 2005b, pp. 84–85, author's translation)

3.2. Simondon's technological vocabulary

While it would be futile (and detrimental) to adopt a linguistic paradigm to understand the technical component of culture, the fact remains that Simondon developed a technological vocabulary in METO. This vocabulary not only integrates technical terms, but also introduces conceptual terms that break with everyday language. He aimed, in fact, to break with common sense, considering the technical object from a utilitarian point of view, and, by extension, to break with all functionalist classifications.

It would take too long to summarize here the development of the classification (see Bontems 2015b, pp. 183–200), which occupies the first third of METO. Let us recall that, on the one hand, Simondon designates different levels of organization of technical objects ("element", "individual", "ensemble", to which one must add the infra-objectal level of the material or terrain and the supra-objectal level of the "network") based on the criterion of recurrence with the "associated environment"; and that, on the other hand, it identifies the technical objects from their "genesis" – the invention of the operation they perform – and studies their evolution ("concretization") within the "technical lineages" toward a "technical essence".

To give a rough idea of the "genetic" classification, here is an example: "engines" is usually used to refer to any technical object that "sets something in motion". In the utilitarian language of common sense, it does not matter how the technical object works; what matters is the function it performs. Simondon observed that, from a technological point of view, an internal combustion engine, an electric motor and a mainspring are unrelated: the mainspring is closer to a crossbow than the other motors. The "engine" class is devoid of technological relevance, whereas the lineage of mechanical engines, whose operation is carried out by decompression of springs, has a genetic reality, both conceptual and material.

The adoption of this vocabulary to designate objects or collections indicates a break with functionalism and even suggests the analyses to be undertaken. On the other hand, it is not sufficient to guarantee the operational value of the classifications obtained. What guarantees the filiations within a lineage is the persistence of the "technical schema": "It is the latency and stability of the concrete schema of organizational invention throughout its successive developments that ground the unity and distinctiveness of a technological lineage" (Simondon 2017, p. 48). As for the continuity of progress, beyond the substitution of one lineage for another, it depends on a "pure schema of functioning" (Simondon 2017, p. 45). Let us leave aside the stakes of the passage between lineages (Bontems 2012, pp. 85–94) to concentrate on the notion of technical schema: what does it correspond to? And what would be an adequate representation of it, since it seems that more than words are needed to describe it?

3.3. For a diagram of the technical lineages

The term "schema" is remarkably polysemic in psychological and philosophical discourse. To grasp its implications in Simondon's work, we refer to the work of our colleague and friend Vincent Beaubois (Beaubois 2015). Let us recall one of his conclusions: Simondon's "technical schema" is not, as in Kant's work, a representation obtained by "the submission of imagination to experience". The schema underlying the technical lineage does not reproduce the structure of a particular object in that lineage. Simondon specifies, from the beginning of METO, that the technical object must be apprehended according to the "temporal sense of its evolution" and not simply in the actuality of "isolated schemas of operation" (Simondon 2017, p. 26). The technical schema is generative. It is the imaginative operation analogous to the technical operation that successive generations of objects perfect. Its input can, of course, be accompanied by images or structural representations, but is not reduced to them. Rather, it refers to the ability to invent new structures to carry out an operation or to identify the operation of a structure:

> The imagination is not simply the faculty of inventing or eliciting representations outside sensation; it is also the capacity of the prediction of qualities that are not practical in certain objects, that are neither directly sensorial not entirely geometric, that relate neither to pure matter nor to pure form, but are at this intermediate level of schemas. (Simondon 2017, pp. 74)

Throughout the research we carried out, in collaboration with engineers, to operationalize the "genetic mechanology"[3], it seemed necessary to us, in order to define the technical lineages that we were studying, to use diagrams to determine the identity of a technical schema. To follow this path, also indicated by Duhem and Beaubois[4], we used the diagram "*Source-cible-flux-champ*" (SCFC, source-target-flow-field) by Jean-Louis Ermine. The latter also proposed a method for drawing "line diagrams" (without limiting himself to technical objects): "It is a tool created and used in the theories that found the history of techniques or technologies of human groups (A. Leroi-Gourhan, J. Baudrillard, A. Moles, G. Simondon: a synthesis of these references in (Deforge 85)). Here it adapts quite well to the context of the evolution of knowledge in general" (Ermine 2000, author's translation).

Here is an example (made with our PhD student Thomas Guy, see Bontems and Guy 2017, pp. 109–120) of an SCFC diagram to define a certain lineage of particle detectors. We studied the genealogy of "Micromegas", a particular type of detector used in particle physics, in collaboration with physicists from the CEA, notably with Ioannis Giomataris, who was its co-inventor with Georges Charpak. In order to identify the schema corresponding to the Micromegas lineage, we borrowed from the MASK knowledge management method (Ermine 1996, p. 118) the diagram schematizing the reaction of a *source* system to a signal that generates a *flow* to a *target* system under certain *field* conditions.

Although very general, this type of diagram is useful to the technologist. It connects Micromegas to the family of *gas ionization detectors* whose internal operations are ionization and amplification (operated, respectively, by fields E1 and E2). This diagram also establishes that Micromegas are, more specifically, related to the lineage of detectors delivering information in the form of an *electrical signal.*

3 "Mechanology" refers to the technological tradition to which Simondon belonged (see Guffroy and Bontems 2018).

4 "[...] the encyclopedism of the Age of Enlightenment paved the way to a visual schematism more adequate to the technical reality" (Duhem 2015, author's translation); "understanding the schematics of an engine or a bridge, on the other hand, cannot be achieved by a simple direct incorporation. It must go through a set of external 'diagrams' which are plans, models, schematics, mock-ups, etc. Simondon did not explicitly develop this theme, but we think it is important to insist on the fact that it is through this means that an affective modality of industrial diagrams, different from that of artisanal diagrams, is made possible: the affective modality here goes through a necessary diagrammatization of the technical diagram, the physical investment of the designer being thus always present although off-center" (Beaubois 2015, author's translation).

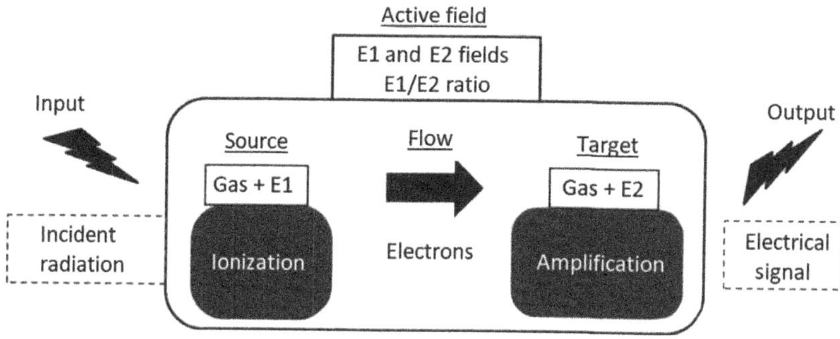

Figure 3.1. *SCFC diagram of the Micromegas*

Our methodology therefore leads to the exclusion of, among other things, fog chambers and bubble chambers, which, despite their similar function, do not deliver information in the form of an electrical signal. The historian Peter Galison has also emphasized the differences between the instrumental tradition of "image-producing instruments" (Galison 1989, pp. 213–251) such as bubble chambers, and that of "logical counting devices" (Galison 1989, pp. 213–251). However, he included in this category not only wire chambers but also spark chambers, even though technologically the internal operations are different. Once the position of the "Micromegas" lineage within the family of gas ionization detectors was defined, we were able to trace its genealogy by identifying the inventions that gave rise to technical lineages that at least partially realized the technical schema of *gas ionization detectors producing an electrical signal*.

3.4. Conclusion

The radicality with which Simondon dismissed the linguistic paradigm and reduced the anthropological scope of language inquiry may well offend the sensibilities of some of our fellow linguists. The very high ambitions of his philosophy of nature in terms of understanding natural and artificial processes also seem, at first glance, to make him an advocate of "ontology" and may appear to many of them as a naive fantasy. However, we believe that the orientations and demands of his research fundamentally converge with those of such a rigorous linguist as François Rastier, who has been committed to dissolving ontological preconceptions of language (Rastier 2015a, pp. 117–149, chapter titled "*Déontologie*") and to denouncing the Heideggerian mysticism of being (Rastier 2015b).

Indeed, by giving absolute priority to "to on", that is to say, to the understanding by the sciences of the physical, biological and psycho-social processes of individuation (as well as the process of concretization) over *logos*, as scholastically conceived by philosophers as the means of the unveiling of being, Simondon actually shattered the usual ontological pretensions of philosophy. We know, for example, how he radically criticized substantialism and hylemorphism, but the similarity of this critique to that of Ferdinand Saussure has perhaps not been sufficiently emphasized: "We tend perpetually to convert by thought into substance the diverse actions that language requires" (Saussure 2002, p. 81, quoted by Rastier 2015a, p. 118, author's translation). Thus, Simondon's critique of ontology is complementary and symmetrical to the critique that the linguist brought to his discipline by rejecting ontological conceptions in order to study the actual structures and operations of language.

Far from denigrating the work of the linguist, the Simondonian philosopher who assumes the demands of rigorous technology must seek his help in articulating his own investigation with the terminology of engineers. This is the meaning of the collaboration we have begun with linguist Marie Calberg-Challot on the lineages of inertial power plants developed by the industrialist Safran. In return, our research to make the "diagrammatization" of technical diagrams and lineages operational undoubtedly suggests new ground for the "science of signs within societies". Rather than dreaming of a pseudo-discipline such as "diagrammatology", the philosopher of technology hopes to take advantage of the achievements of the scientific disciplines already established and that they will take into account his own technological efforts by taking up the diagrammatic question in order to shed light on his own approach.

3.5. References

Beaubois, V. (2015). Un schématisme pratique de l'imagination. *Appareil*, 16 [Online]. Available at: http://journals.openedition.org/appareil/2247 [Accessed 19 September 2018].

Bontems, V. (2012). Simondon, le progrès et l'évolution des lignées techniques. In *Formes, Systèmes et Milieux techniques (après Simondon)*, Parrochia, D., Tirloni, V. (eds). Jacques André Editeur, Lyon, 85–94.

Bontems, V. (2015a). Le cycle de la technologie chez Simondon. In *Cahiers Simondon 6*, Barthélémy, J.-H. (ed.). L'Harmattan, Paris, 7–27.

Bontems, V. (2015b). Sur la classification des machines selon Simondon. *Artefact*, 3, 183–200.

Bontems, V., Guffroy, Y. (2018). La mécanologie: une lignée technologique francophone. *Artefact*, 8, 255–280.

Bontems, V., Guy, T. (2017). L'étude des lignées phénoménotechniques. De Bachelard à Simondon et aux Micromégas. In *Il senso della technica. Saggi su Bachelard*, Donatiello, P., Galofaro, F., Ienna, G. (eds). Esculapio, Bologna, 109–120.

Carrozzini, G. (2015). La contribution de Gilbert Simondon au naturalisme. *Appareil*, 16 [Online]. Available at: http://journals.openedition.org/appareil/2206 [Accessed 19 September 2018].

Duhem, L. (2015). Simondon et le langage. *Appareil*, 16 [Online]. Available at: http://journals.openedition.org/appareil/2223 [Accessed 19 September 2018].

Ermine, J.-L. (1996). MKSM : méthode pour la gestion des connaissances. *Ingénierie des systèmes d'information*, 4(4), 541–575.

Ermine, J.-L. (2000). *Les systèmes de connaissance*. Hermes, Paris.

Galison, P. (1989). Bubbles, sparks, and the postwar laboratory. In *Pions to Quarks: Particle Physics in the 1950s*, Brown, L., Dresden, M., Hoddeson, L. (eds). Cambridge University Press, Cambridge, 213–251.

Rastier, F. (2015a). *Saussure au futur*. Les Belles Lettres, Paris.

Rastier, F. (2015b). *Naufrage d'un prophète*. PUF, Paris.

Ricœur, P. (1960). *Être, essence et substance chez Platon et Aristote : cours professé à l'université de Strasbourg en 1953-1954*. Le Seuil, Paris.

de Saussure, F. (2002). *Écrits de linguistique générale*. Gallimard, Paris.

Simondon, G. (1960). Forme, information, potentiel. *Bulletin de la Société Française de Philosophie*, 54(4), 143–174.

Simondon, G. (2005a). *L'individuation à la lumière des notions de forme et d'information*. Millon, Paris.

Simondon, G. (2005b). *L'invention dans les techniques. Cours et conférences*. Le Seuil, Paris.

Simondon, G. (2010). *Communication et information. Cours et conférences*. La Transparence, Paris.

Simondon, G. (ed.) (2014). Culture et technique. In *Sur la technique (1953-1983)*. PUF, Paris, 315–330.

Simondon, G. (2017). *On the Mode of Existence of Technical Objects*. Translated by Malaspina, C. and Rogove, J. University of Minnesota Press, Minneapolis.

PART 2

Hermeneutics of Science, Hermeneutical Sciences

4

On the Philosophy of Mathematics: Reflections on "Making Science", Based on Cavaillès

4.1. Mathematics, a precondition of rational philosophy

Why is there a need for this discussion on the philosophy of mathematics, in a framework that seeks above all to celebrate the thinking and cultural sciences? At first glance, this may seem at odds with the purpose of this book, which, in the general field of science, is concerned with what falls within the domain of the human sciences.

In fact, it is a question of reflecting on the question of the academic territorialization of knowledge, which in many respects seems to be non-congruent with the cultural and scientific spaces actually occupied, saturated, or only crossed by this knowledge. The mathematical science curriculum is of course not reducible to the postulate of a science that aims at accuracy.

What can the philosophy of mathematics bring to the thematic perspective *Making Sense, Making Science*? Perhaps first of all the fact that the heuristic orientation of the field of philosophy of mathematics is already the guarantee of a fruitful disciplinary diversity. What does the philosophy of mathematics do? It is an offshoot of the philosophy of science, and it mainly questions the foundation of mathematics, i.e. its efficiency, its necessity, the degree of existence of the entities that constitute mathematics, the nature of mathematical objects and also the truth in it.

Chapter written by Franck NEVEU.

For this brief overview of the question, we will start from the common point between philosophy and mathematics, agreeing that mathematics combines intuition and evidence in a singular way, which is also what the philosophical text does.

On the question of the separation of mathematics and philosophy, we can undoubtedly recall with Alain Badiou (Badiou and Haéri 2015), that this separation, as it is known to us today, would undoubtedly have greatly astonished the classics, since many philosophers of the 17th and 18th Centuries were also great mathematicians.

Badiou recalls in particular that Descartes was a founding mathematician, creator of analytical geometry, unifying geometry and algebra (he showed how a curve in space, therefore a geometric object, can be represented by an equation) (Descartes 2009). Leibniz (1989) was also a mathematician, founder of modern differential and integral calculus. Generally speaking, the history of philosophy teaches us that mathematics and philosophy have linked origins. Plato, Descartes, Spinoza and Kant, even Searle, formally declared that if there had been no mathematics, there would have been no philosophy. In other words, mathematics was conceived very early on, and this is explicit in Plato's (1940, 1942) work (*Meno*, on virtue; *Parmenides*, on ideas; and *Timaeus*, on the origin and nature of the physical world), as a precondition for rational philosophy to be born, because mathematics is the example of a process of knowledge that holds by itself and of itself.

In antiquity, mathematics constituted a universe where ideas considered as true, were demonstrable and circulated under the condition of their validation. With the mathematician, in antiquity, we were dealing with a universality totally freed from any mythological or religious presupposition, from any judgment of authority. The narrative was replaced by evidence. As historians of philosophy have shown, mathematics challenges traditional narratives. Evidence depends only on rational demonstration, and it is refutable in principle. In this sense, as Badiou points out, mathematics is part of democratic thought, and philosophy has only been able to constitute itself autonomously from the religious narrative with the formal support of mathematics. Evidence is the natural authority of reason.

It is therefore a fact that there is from the outset an interlinked part between mathematics, democracy (in the sense of political modernity as opposed to traditional authorities) and philosophy. The thoughts of Parmenides can even be mentioned (Dumont 1990), which predate Socrates and Plato and date back to the 5th-Century BCE, as a corpus in which one finds traces of methods of thought that find their highest fulfillment in mathematics.

4.2. Reasoning by the absurd and excluded middle

In his seminar in 1985, Badiou (2014) developed the example of reasoning by the absurd that can be considered decisive for the rational power of mathematics. According to absurd reasoning, one proves that a statement p is true, not by directly constructing its truth from already established truths, but by demonstrating that its contradictory statement (non-p) must be false. This is the application of the law of the excluded middle. Given p, a well-formed statement, i.e. one that conforms to the syntactic rules of the system under consideration, either p is true or non-p is true. There is no third way. A truth is thus established from a false hypothesis.

How do we prove that non-p is false? By assuming that it is true, and by drawing consequences from this assertion that contradict already established truths. We then apply the principle of non-contradiction: since non-p contradicts a statement (for example q), which is true, and two contradictory statements cannot be true, then non-p must be false. And therefore, p must be true.

> You see the amazing path of the evidence: you want to establish that p is true, for which you have your reasons (this is your hypothesis). To this end, you fabricate the fiction "non-p is true", which you hope is false! And to nourish your hope, you draw consequences from this fiction, thus moving with implacable logic into what you think is false, until you encounter a consequence that explicitly contradicts a previously proven true statement. This controlled, regulated navigation between true and false is, in my opinion, quite characteristic of nascent mathematics, of the division that it introduces with any revealed truth or whose force would only be poetic. This "tone" can be found in Parmenides. And we find it because, in order to prove that being is the primary truth, he establishes that non-being is not. He therefore reasons by the absurd. My conclusion is clear: rational philosophy and mathematics were born at the same time, and it could not be otherwise. (Badiou and Haéri 2015, pp. 33–34, author's translation)

Henri Cartan, exposing the question of the foundation of mathematics in Cavaillès' work, did not say anything else, but differed:

> The fact that some propositions may be neither true nor false in a given theory leads us to say a few words about the famous quarrel of the excluded middle. The law of the excluded middle, which dates back to antiquity and was used in particular by Euclid, consists of the following: to prove that a proposition A is true, one provisionally adds to the axioms of the theory the negation of A (by saying "suppose that A

> is false"); and if one arrives at a contradiction in this new theory, one considers that one has proved that A is true in the initial theory! This is the "demonstration by the absurd". This mode of reasoning is illegitimate, intuitionists say, because you have assumed that A was either true or false, which is generally not the case. Not at all, reply the formalists: this reasoning by the absurd finally amounts to proving the truth of the following proposition: no A => A (the negation of A leads to A) and this proposition, according to the rules of logic, is rigorously equivalent to the proposition A itself. In other words, any demonstration by the absurd can always be put in the form of a demonstration that avoids the use of the absurd. In fact, I know of no mathematician who today rejects this law of the "excluded middle". (Cartan 1984, author's translation)

In the end, Descartes did not proceed otherwise and retained mathematics in his philosophical undertaking as the ideal of demonstration. It was well known to Descartes that the philosophical text must take the form of a long chain of reasons, which constitute the very essence of mathematics. But he also used reasoning by the absurd. In order to prove the existence of an entity (for example the existence of the outside world), he did not proceed directly but developed the fiction of an absolute doubt that would be tantamount to affirming the nothingness of all truth and all experience. And then he realized that the very fact of doubt could not be doubted. It is the *cogito* that establishes a point of truth (the "I exist") by negating the absolute negation that is doubt. It is the same approach to proving the existence of God. From the certainty that I have an idea of the infinite, while I am finite, results in the necessary existence of an infinite being who has created this idea in me.

4.3. The final causes

We could further develop, as Badiou did, with Spinoza, who discussed *Ethics* (*Ethics*, I), by positing the idea that without mathematics humans would have remained in ignorance, among other things because we would have continued to explain everything through final causes, mythologies and supernatural powers. *Ethics* for Spinoza was thus inscribed in the idea that it is a possible consequence of the existence of mathematics. The crucial role of mathematics for Spinoza was precisely to have discredited explanations by final causes, in other words, to have banished finality from the philosophical field and to stick to deductive sequences.

Spinoza recognized three kinds of knowledge: a mixture of sensitive representation and imagination, which is ordinary ignorance; ordered conceptual knowledge, the point-by-point demonstration, of which mathematics offers the paradigm; and intuitive attendance of God, which is the name of nature or totality, which is properly philosophical knowledge. But Spinoza clearly demonstrated that without access to the second, there can be no question of accessing the third.

And we could go even further with Kant, who reminds us in the introduction to *Critique of Pure Reason* of the absolute necessity of mathematics for philosophy to exist, and particularly critical philosophy. Badiou reminds us of the fundamental question asked by Kant "where does the universality of science come from?", a question that would be irrelevant if there were no science. And without mathematics there is no natural science.

4.4. "Universally true" judgments

We know the two possible conceptual approaches to mathematics that are widely discussed in every sense in the philosophy of science. A Platonic, realistic conception, according to which the object of mathematics exists outside humankind. A formalistic conception, according to which mathematics is a pure creation, in particular because it involves the creation of a specific formal language. Aristotle considered mathematics as an esthetic, because he did not see any relation with reality, he considered it as an arbitrary creation producing a certain intellectual satisfaction. Plato, however, saw it as the foundation of universal rational knowledge, which is why all philosophy must begin with mathematics.

Kant constructed a rudimentary conception developing the theory of an organization of mathematical thought that would not come from concrete experience but would be prior to it because it exists with regard to experience *a priori* and not *a posteriori*. As a result, what is at stake in the formal sciences is the subjective organization of human knowledge, in other words, what Kant calls the transcendental subject. So rationality is universal, but it is not universal because it touches reality, it is universal because it refers to a universal structure of cognitive subjectivity itself. Thus, if we agree on the value of a mathematical demonstration, it is not because it says something true about the world or because it affects the real world, but because the human intellectual structure obeys a single paradigm.

For Kant, mathematics is not universal because it thinks of the formal structures of a being as being but because it exists through a language coded in the same way for everyone. Of course, for Kant as for Descartes or Spinoza, from the moment mathematics was invented by Thales (it is Kant's theory that it is due to the singular genius of the Greek philosopher who invented demonstrative geometry and arithmetic), it paves the way for the development of science.

Without mathematics, the tired but sensible philosophical question of *where universally true judgments come from* could not be formed. Mathematics is the position of many "Platonic" philosophers, for the science of being, and in this it is crucial in philosophy. This is the meaning of the Platonist adage on the pediment of the Academy: "Let no one ignorant of geometry enter here", *here* probably referring to the place of thought, of philosophy.

4.5. The linguistic problem of mathematics

One of the most crucial and interesting questions for us is that of the linguistic problem of mathematics, which escapes the singularity of languages, even if of course languages are used for the exposition of mathematical knowledge (they transmit it, but they are not that language). Mathematical language does exist as a specific language, not natural, but formal and conventional. However, one of the persistent philosophical questions that have been raised throughout history is that of the link between natural language, which is singular, and rational discourse with a universal vocation. Is there not a possible interference, a risk of deformability, an effect of illusion of universality in the fact that a rational discourse of universal scope is held in a singular language?

We know that Descartes developed the theory of the effability of *reason*, and that for him reason could not be thwarted by the singularity of languages (Derrida 1990; Descartes 2009). The fact is that mathematics, as Badiou points out, is interested in the most formal and abstract dimension of being as being, but from the principle that everything that exists is composed of a multiplicity. As a result, mathematics returns to the idea of a general theory of the different forms in which multiplicities acquire a certain consistency.

It is not the theory of what is as it is this or that, but only as it is. With this peculiarity that it is not a question of thinking about the totality of the relationship of the subjects with the world.

> Mathematics is not at all the science of the difference between a fall foliage and a summer sky; it only means that all these are multiplicities, forms that have something in common, the fact of being, quite simply. And it is the abstract forms of this "common" that mathematics tries to think about. (Badiou 2015, p. 43, author's translation)

Badiou saw the separation between philosophy and mathematics begin at the end of the 19th Century and in the first half of the 20th Century, with the development of an "anti-philosophical" current (represented in particular by Nietzsche, Wittgenstein), which modified the program of philosophy in a perspective that had not been his since Plato, in particular by developing the idea that one had to abandon the program of assuming the complete and systematic character of philosophy, which made this separation possible and undoubtedly precipitated.

A separation, therefore, whose reasons were historical, Hegel's philosophical romanticism and Sartrean existentialism, in particular, contributed to moving philosophy away from analytical and demonstrative rationality. From the French Revolution onwards, the concern for history had valued movements, revolutions, negativity, at the expense of "the species of contemplation *sub specie aeternitatis* of mathematical truths", which become timeless as soon as they were established. A separation also for the sociological and the institutional, i.e. academic, reasons linked to the artificial territorialization of knowledge, in particular the segmentation between literary and scientific knowledge into radically disjointed groups.

4.6. The epistemological break: the explanatory versus comprehensive method

This brings us to the preamble, and no doubt also to the heart of *Making Sense, Making Science*'s argument, namely the compartmentalization, the "separation" between two types of cultures that are presumed to be different or even incompatible: literary and scientific culture.

On the reasoned emergence of these two approaches, one can only refer to Wilhelm Dilthey and his 1883 publication *Einleitung in die Geisteswissenschaften* (translation: Introduction to the humanities) (Dilthey 1992), where he takes a stand against the Comte's positivism, whose theories are distorted at least by the fact that there is not one method of knowledge, but two: one is the natural sciences and the other is the humanities or the sciences of mind.

For Dilthey, what prevails in the natural sciences is the explanatory (thus causalist and nomological) method. Here, we study phenomena whose intelligibility is extrinsic because it depends only on the regular relations that these phenomena have with other phenomena of which there are the effects.

What prevails in the sciences of mind is the comprehensive method, because these sciences take as their object the results of actions whose intelligibility is intrinsic. These actions are not explained by efficient causes external to the actors, but by subjective and conscious motivations because of which they act: "The game of acting causes given to us in nature is here replaced by the game of motives and ends" (Dilthey 1992, author's translation).

Weber (1965), in *Essais sur la théorie de la science* (translation: Essays on the theory of science), stated that this theory has the merit of underlining the error of scientists and believed that the sciences of mind must and can imitate the method of the natural sciences.

Contrary to Comte, Durkheim and the entire positivist tradition, Weber opposed the idea that to be worthy of the name of science, sociology had to imitate physics by adopting the postulate that a phenomenon can only be the effect of its efficient cause, i.e. an antecedent phenomenon of the same nature.

Scientific judgments about natural phenomena are often counterintuitive. It takes more than an effort to stop believing that the sun revolves around the earth. Similarly, spontaneous physics judges that mass is what is big or heavy. Since Newton, physicists have known that mass is the quotient of force and acceleration. And this relationship is by no means intuitive, and it is only understandable if we break with the naive physics of fat and heavy which, as Bachelard reminds us, is an obstacle to the intelligence of the concept of mass.

Returning to Dilthey and the opposition between explanation and comprehension, the dividing line can be said to be constituted by the historicity of objects specific to the sciences of mind (on the explain/understand opposition, see Colliot-Thélène 2004, pp. 6–23).

Dilthey undoubtedly owes Hegel the consideration of the historical dimension of thought. But he takes up from Kant's rejection of all metaphysical speculation and extends his critical enterprise to the sciences of mind, whose objectivity must be defined and delimited, and his *Einleitung in die Geisteswissenschaften* (Introduction to the humanities), which he presents as a "critique of historical reason", aiming to establish the specific laws of social, moral and intellectual facts that characterize cultural systems or social organizations. Its objective is to escape the apriorism of reason and the empiricism of history.

4.7. The understanding

This is undoubtedly a possible point of contact with Jean Cavaillès, not because he was inspired by Dilthey, but because he also proposed this dual constraint. From a methodical point of view, explanation for Dilthey involves a purely intellectual process, whereas understanding mobilizes all the capacities of the mind applying itself to interpreting the productions. There is a correspondence between the historicity of the object to be understood and the method of understanding it in a circularity that destroys the representation of the object by a subject. The knowledge of history is history itself, Dilthey said, and this rather Hegelian formula was taken up by Cavaillès in order to understand the founding status he gives to the historical method in mathematics by distinction with the logical method (notably that of Frege and logicists).

Dilthey values historical experience as a specifically human way of knowing, the way of knowing the truth, as distinct from the absolute truth of traditional metaphysics, from Kantian apriorism, from scientific positivism, and from the individual subjectivity of the introspective method. To understand is to interpret at different levels and in different kinds of contexts in order to highlight stable temporal formations showing systemic relationships indicative of an objective meaning. To understand is to grasp internal relations that as a whole show coherence or identity, whereas to explain establishes an external and segmental relationship of cause and effect.

Cavaillès worked a lot on "understanding", which structured his philosophical approach. The theme is not reserved for the "human sciences". Mathematics must take hold of the concept. This was the case for Cavaillès in terms of rational knowledge, and he did not take up the Kantian division between philosophical and mathematical knowledge. Intuition does not divide mathematics and philosophy. But the division between reason and understanding according to Spinoza – for whom the essence of reason is to understand – is to artificially grasp what the understanding discerns or distinguishes.

Defining *philosophy* by *understanding*, while rejecting the connection between philosophy and the human sciences, which are built on facts and their narrative, is a line of action in Cavaillès' thinking. In this, he went against the tide of his time and the renewed hermeneutics that characterized the beginning of the 20th Century, whether it was Husserl's logical research or Heidegger's Dasein philosophy, based on a theory of the unveiling of truth foreign to scientific reason and close to the religious exegesis of revelation and testimony, as Ricoeur would later use.

Cavaillès reserved understanding for philosophy and radically distinguished it from description and narrative. Philosophizing for Cavaillès was a third path between the exact and experimental sciences and the humanities, like Dilthey. To philosophize is neither to describe nor to testify, it is to understand from within. But this does not mean that it refers to a subjective interiority. It is to head toward what is necessarily what it is, to understand from within, it is to understand the essence, the internal necessity, to reach objectivity in its essence in a central intuition.

Mathematics is not a text to be interpreted, it is in fact a networked development, a moving web whose nodes must be found and from each of them the weave must be extended. There is nothing to interpret, but everything to understand, i.e. to do.

4.8. Mathematics as becoming

For Cavaillès to know is not to know for sure nor to deduce without fault, it is to wager on the success of certain steps. "To know the world is to wager – to wager that certain actions will succeed [...]" (Cavaillès 1994, p. 650, author's translation). This is precisely what the best specialist in Cavaillès' philosophy says:

> Cavaillès is a philosopher of evidence, of thought united with action, of "thought in action", as he wrote, of thought as experience, and not just thought of experience. For a "militant" mathematician, mathematics is not or not only an object of statement, discourse and deduction, but a matter to be worked on, to be transformed. Mathematical gestures are acts of thinking on objects of thought [...]. But, conversely, an object of thought, a number, an operation (adding 3 and 4, extracting the square root of 9), a geometric property (being equilateral for example), is the product of a sequence of mathematical gestures. The mathematical object is internal to mathematical thinking, matter is coextensive with form. (Benis Sinaceur 2013, p. 75, author's translation)

At the end of his last posthumous work, *Sur la logique et la théorie de la science* (in English, *On Logic and the Theory of Science*), Cavaillès developed this epistemological perspective:

> [...] one of the essential problems of the doctrine of science is that [...] progress is not an increase in volume by juxtaposition, the former remaining with the new, but a perpetual revision of the contents by deepening and erasure. What is after is more than what was before, not because it contains it or even prolongs it, but because it necessarily emerges from it and bears in its content the singular mark of its superiority each time. There is more consciousness in it – and it is not the same consciousness. The term consciousness does not have a univocity of application – nor does it have an isolable unity. There is not a consciousness generating its products, or simply immanent to them, but it is each time in the immediacy of the idea, lost in it and lost with it, and linking itself with other consciousnesses (what one would be tempted to call other moments of consciousness) only by the internal links of the ideas to which they belong. Progress is material or between singular essences, its driving force, the need to surpass each one of them. It is not a philosophy of consciousness but a philosophy of the concept that can give a doctrine of science. The generating necessity is not that of an activity, but of a dialectic. (Cavaillès 1994, p. 560, author's translation of the original French edition)

Mathematics is therefore in essence a becoming. According to Cavaillès, it is a paradoxical becoming because it is autonomous. The history of mathematics is not strictly speaking a history since it took place in a temporality of its own, marked by necessity. The self-generation of concepts and structures runs throughout history.

As Pierre Cassou-Noguès (2017) points out, this story is unpredictable and its necessity escapes the grip of logic. It is the result of a dialectic, which consists of the generation of new theories by "tearing" and restructuring the mathematical fabric through the formation of new concepts, which dissociate, redefine themselves and establish unpredictable links between them. Here, we break with the idea of a linear enlargement by juxtaposing knowledge, which would make it possible to predict the future state of mathematics from the present state. What Cavaillès said is that although in retrospect it is intelligible, its future necessity is unpredictable. Mathematics develops by perpetual reorganization and separates while integrating previous knowledge, open to the future.

Thus, to draw a philosophy of mathematics from an analysis of their history, it would be necessary to show that their history, that is to say for Cavaillès their becoming, is not extrinsic to mathematics but gives a means to access their reality. Cavaillès went beyond this stage and, for him, a becoming rather than a path to mathematical reality is mathematical reality. This cannot be conceived as a system of objects and relations in itself, immutable and which mathematics would discover region by region. This immutable reality would then have no immediate connection with the theories on which mathematicians work. These theories are constantly changing and are never fixed definitively.

In fact, new theories lead to reworking old ones. Cavaillès realized this by following the work of Cantor and Dedekind. Even arithmetic is caught in this future since the theory of sets with the new operations it allows defines calls for a reformulation. And this is what is to be understood from what is quoted above ("One of the essential problems of the doctrine of science [...]"). In other words, nothing is subtracted from the becoming in the mathematical edifice. The current theories represent only a transitory state that must be put back into the mathematical becoming. Becoming is the movement in which the theories make sense. Reality is not part of a theory in becoming, but in the very becoming of the theories.

Mathematics is, rather than the actual field, the becoming that crosses and binds the successive fields. This is what is to be understood from Cavaillès' statement: "Mathematics is a becoming". Mathematical theories, never fixed, are always subject to a possible revision and therefore caught in a becoming. We cannot situate mathematical reality in objects in themselves. We cannot situate it in current theories, which are only transitory. Rather, it must be placed in their very becoming.

4.9. Truth and metalanguage

Returning to the problem of the territorialization of knowledge and the epistemological question, what catches my attention as a linguist is the parallel that can be drawn between disciplines, or even epistemological analogies. This is the antithesis of the epistemological theories of Thomas Kuhn (1972). As we know, Kuhn's theory is based on two fundamental concepts: the concept of paradigm, i.e. a scientific current that has dominated for a fairly long period and which can be defined as "normal" science, and the concept of paradigm shift, corresponding to a scientific revolution leading to the replacement of one paradigm by another. Kuhn's theory contrasts with that of Popper (1973), for whom the progressive evolution of disciplines takes place in frequent leaps and bounds and resembles a permanent revolution.

The Kuhnian model was discussed in linguistics in the 1980s, particularly among language science historians, and the prevailing view at the time was that this model could be applied to the field of linguistics. The evolution of scientific work in this field today suggests that this is not the case, not least because linguistics is a discipline with a low rate of re-enrolment. This re-enrolment rate notion is used to measure a discipline's ability to integrate its knowledge. If there are many theoretical breaks in a discipline and therefore a low re-enrolment rate, the previous states of the discipline remain of direct theoretical interest. This is the case in linguistics where it is often necessary to take into account the previous states of descriptions and analyses. There are many examples of this (the *modi signifandi* "modes of signifying") in medieval theories, which make it possible to dissociate words/forms/functions or, according to Beauzée, the encyclopedist grammarian, who in his study of the genitive reveals a most innovative approach to the problem of determination and reference, where for example it is shown that a proper noun can play the adjective role in the nominal syntagm, which extends the notion of adjectivity to a perimeter infinitely wider than that of the parts of discourse.

At the end of the day, it appears that in two fields as distinct and seemingly as divided as mathematics and linguistics, one supposedly falling within what is improperly called the *exact sciences*, the other falling within what is equally improperly called the *human sciences*, one operates on and with categories of thought. The notions and entities they have to deal with fall, in both cases, within what Dilthey called the "sciences of mind". As such, without neglecting the dissimilarities in scientific objectives and observables between the two fields, the radical division between the fields can be taken for granted and not proven, if of course comparability between the scientific fields is considered efficient. Why is this so? We must ask ourselves this question, and from a philosophical point of view, a good angle of attack would be that of truth and metalanguage.

As Jean-Claude Milner reminds us, one cannot say of anything that it is true or not unless it is itself of the order of a possible saying (Milner 2014): a proposal that can be said, a thought as it could be said, a state of affairs as it is considered to be transmissible by a saying, etc. (Milner 2014). Since there is only truth in what is said and since there is only truth if one can say the "truthfulness" of what is said, the truth is always "truth of". Truth, as it would not be truth of, does not exist. Since there is truth only in what is said, and since there is truth only if one can say the "truthfulness" of what is said, there is truth only if there is metalanguage. There is no internal difference between what is untrue and what is true. This fact allows grammar, as it differs from logic: a false sentence can be correct, a true sentence can be incorrect. But logic itself is not enough: a logically well-formed proposition can be false. As a result, Milner continues, the predicative definition of truth and the doctrine of correspondence jointly appear. The doctrine of correspondence and metalanguage also jointly appear. If, however, there is no metalanguage if it is impossible for one person to say anything about another person's words, then the truth is not said.

Beyond this problem of truth and metalanguage, which could allow a closer relationship than one might imagine between these two fields of knowledge, which are so dissimilar, it must be acknowledged that a dividing line seems to be drawn which could re-actualize the epistemological division developed by Dilthey around the concept of observation.

4.10. The theoretical in difficulty, an aspect of the epistemological shift in linguistics

In recent decades, the development of digital technology, documentary informatics and so-called "corpus linguistics" has profoundly changed our relationship with the empirical in language sciences. The spectrum of facts to be studied as well as their nature has changed significantly. New modes of observation associated with new modes of explanation have given rise to new realities.

The theoretical is affected in a very significant way, in particular because the "principles" must be "applicable" in the analysis, so that the theories from which these principles are derived are constantly undergoing the test of deformability. The theoretical principle must be modifiable at will according to what is sometimes too hastily accepted as being the reality of language or discourse simply because it comes from a material that is self-given, and one accepts without too much discussion on the relevance of it. This is, in a way, the price to be paid for the phenomenal development of the linguistics of usage. Thus, the science of language has had to make an epistemological shift, passing, one might say, from the *exemplum* to the *datum*. The consequences are not negligible, and they are both positive and negative. The advantages appear in the necessary development of a reflexive point of view within the methodologies, which marks a break with a certain form of candid objectivism that has long prevailed, an objectivism according to which the sciences would be determined solely by their internal logic, with no regard for the formal and material structure of their discourse. Paradoxically, language sciences have largely developed this perspective, whereas one might have expected this field to have a better understanding of the effect of discourse on scientific ideas. Another advantage of this epistemological shift can be found in the interdisciplinary connections that have become apparent. State territories have abandoned their implicit preconceptions and have in a way become de-ideologized for the benefit of science. As for the disadvantages of moving from the *exemplum* to the *datum*, they are clearly apparent in the false idea that the factual could be seen and grasped without prior definitional work. Empirical naivety has taken the place of theoretical naivety, and the infinite availability of language materials, made possible by technological development, suggests that this naivety will undoubtedly be more lasting and more pernicious. The growing role of computer science in this linguistics of usage produces illusionary effects on the data, simply because of the ever-increasing technical performance of the tool. Thus, we frequently shift from a scientific objective to a technical objective, forgetting in the process the imperatives of linguistic research. The excess of empiricity and the lack of generality necessary for theory can be thought of as a denial of scientificity. In relation to this state of affairs, linguists are the actors, lucid or not, of a form of in-depth consolidation of their scientific field.

This "new" conception of what is meant by "doing-science" in linguistics would undeniably benefit from taking into account the question of the impact of the interpreting subject in the chain of the analytical process, of which the *term* (the scientific term) is an essential link (the semiotics of terms in linguistics remains an abandoned project). As Nicolaï (2007) has shown in his work on the "vision of facts" in language science, the identification of a linguistic fact presupposes a prior agreement on the relevance of the identification itself. There can be no possible identification without the pre-existence of a paradigm, in the Kuhnian sense, i.e. an intersubjective agreement both on the form and on the potential value of the fact, or even on the methodology that enables it to be grasped. In the scientific field, the interpretative act is not only in the *terminus ad quem*, but also in the *terminus a quo*. It is in the sequencing of the "data", in the construction of the facts, in the theoretical framework, in the selection of relevance and generally speaking in the entire analysis *a priori*. The interpretative act as the "rudimentary" reality of the analysis is observed in decoy effects, i.e. false evidence, which will determine in particular the analogical approaches leading to the identification of formal or functional relatives. The recognition of a structural isomorphism necessarily presupposes the relevance of a comparative approach.

Knowing how to take into account the researcher's false absence in the analysis implies never abandoning reflective thinking. This also makes it possible, therefore, to maintain in a state of awakening the hermeneutic attention necessary for scientific activity.

4.11. References

Bachelard, G. (1951). *L'activité rationaliste de la physique contemporaine*. PUF, Paris.

Badiou, A. (2014). *Le Séminaire – Parménide. L'être 1 – Figure ontologique (1985–1986)*. Fayard, Paris.

Badiou, A., Haéri, G. (2015). *Éloge des mathématiques*. Flammarion, Paris.

Benis Sinaceur, H. (2013). *Cavaillès*. Belles Lettres, Paris.

Cartan, H. (1985). Cavaillès et le fondement des mathématiques. In *Jean Cavaillès, philosophe, résistant*. CNDP, Amiens, 15–19.

Cassou-Noguès, P. (2017). *Un laboratoire philosophique. Cavaillès et l'épistémologie en France*. Vrin, Paris.

Cavaillès, J. (1994). *Œuvres complètes de philosophie des sciences*. Hermann, Paris.

Colliot-Thélène, C. (2004). Expliquer/comprendre : relecture d'une controverse. *Espace Temps. L'opération épistémologique. Réfléchir les sciences sociales*, 84–86, 6–23.

Derrida, J. (1990). *Du droit à la philosophie*. Galilée, Paris.

Descartes, R. (2009). La géométrie. In *Œuvres complètes, III, Discours de la méthode* et *Essais*, Beyssade, J.M. (ed.). Paris, Gallimard.

Dilthey, W. (ed.) (1992). Einleitung in die Geisteswissenschaften. In *Œuvres 1. Critique de la raison historique. Introduction aux sciences de l'esprit*. Le Cerf, Paris.

Dumont, J.-P. (ed.). (1988). *Les présocratiques*. Gallimard, Paris.

Hegel, G.W.F. (2006). *Phénoménologie de l'esprit*. Vrin, Paris.

Husserl, E. (1961). *Prolégomènes à la logique pure*. PUF, Paris.

Kant, E. (1975). *Critique de la raison pure*. PUF, Paris.

Kuhn, T. (1972). *La Structure des révolutions scientifiques*. Flammarion, Paris.

Leibniz, G.W. (1989). Nova methodus pro maximis et minimis, itemque tangentibus. In *La naissance du calcul différentiel*, Blay, M., Sinaceur, H. (eds). Vrin, Paris.

Milner, J.-C. (2014). *L'Universel en éclats. Court traité politique 3*. Verdier, Lagrasse.

Nicolaï, R. (2007). *La Vision des faits. De l'a posteriori à l'a priori dans la saisie des langues*. L'Harmattan, Paris.

Plato. (1940). *Œuvres complètes*. Gallimard, Paris.

Plato. (1942). *Œuvres complètes*. Gallimard, Paris.

Popper, K. (1973). *La logique de la découverte scientifique*. Payot, Paris.

Ricœur, P. (1986). *Du texte à l'action. Essais d'herméneutique II*. Le Seuil, Paris.

Spinoza, B. (1977). *Ethica Ordine Geometrico Demonstrata, 1677. L'Éthique*. Vrin, Paris.

Weber, M. (1965). *Essais sur la théorie de la science*. Plon, Paris.

Wittgenstein, L. (2005). *Investigations philosophiques*. Gallimard, Paris.

5

The Semiotic Articulation of Textual Meaning: Significance, Signification, Designation and Expression

What can be said about textual meaning by a text-based semantics that is epistemologically founded upon the semiotic character of language resources? It is known that Benveniste, in his elaboration of the distinction between *semiotic* and *semantic* levels, mentioned that the articulation of the semioticity of linguistic units with the thematico-intentional, praxeological and hermeneutic dimensions of meaning is problematic to the point that the "hiatus" between these two levels is justification for him to envisage the existence of two distinct linguistics, each conceptually matched in a specific way[1]. As it appears to share the same position, a whole tradition of phenomenology of language, although sensitive to the semiotic articulation of languages, insisted on the solution of continuity between meaning and linguistic semioticity, supporting this observation with the fact that the signifier is

---

Chapter written by Régis MISSIRE.

1 "It is a question of whether and how we can move from the sign to the 'word'. In reality, the world of the sign is closed. From the sign to the sentence there is no transition, neither by syntagmation nor otherwise. A hiatus separates them. It must therefore be admitted that language comprises two distinct domains, each of which requires its own conceptual apparatus. For the one we call semiotics, the Saussurean theory of the linguistic sign will serve as a basis for research. The semantic domain, on the other hand, must be recognized as separate. It will need a new apparatus of concepts and definitions. [...] The semiology of language has been blocked, paradoxically, by the very instrument that created it: the sign. The idea of the linguistic sign could not be discarded without suppressing the most important character of language; nor could it be extended to the entire discourse without contradicting its definition as a minimal unit." (Benveniste 1974, pp. 65–66).

fading away in favor of the meaning of speech, both in reception[2] and in production[3]. In other currents of linguistics, however, proposals have been made to try to carry out the analysis in its entirety within the semiotic framework. Coseriu (1997) thus put forward a "textual sign" model in his linguistics of texts that encompasses three levels (*signification*, *designation*, *meaning*), the last being considered as the *signified* of the first two, which together play the *signifier* function of the textual sign:

> [...] Between designation, the signified and meaning, the relation is as follows: designation and signified, that is to say, what the signs denominate, taken together, constitute, in the text, the expression of a higher unit of content, of a more complex nature: the expression of meaning. In a manner analogous to the Saussurian distinction between *signifier* and *signified*, which is valid for the linguistic sign, we also distinguish, in the case of the textual sign, between *signifier* and *signified*: signified and designated together constitute the *signifier*, while the meaning constitutes the signified of the textual sign [...]. Linguistic signs have a signified through which they designate something extralinguistic. This state of affairs in turn represents, on a higher semiotic level, the expression of a unit of content that is of a higher nature, the meaning [...] What is designated in a text, particularly a literary text, in turn becomes expression, the symbol of a

2 "When faced with the spoken chain, we hear meaning, and meaning is again what we pronounce as our mouth articulates sounds: such is our experience of words. This is because of the linguistic signifier's ability to become completely transparent in favor of the signified – except of course when the opposite intent, namely to highlight the signifier, drives the order of the message, as is the case in the use that art makes of language [...]. But the condition of everyday language – whose tenets are communication and economy of means – is the erasure of phonic matter in favor of signification, that is, the signifier's transparency. [...] This limpidity that allows the signifier, in the experience of articulated language, to disappear almost completely behind meaning is not easy to grasp: I invoke a meaning, words form in my mouth, and the interlocutor hears a meaning." (Lyotard 2011, pp. 78–79).

3 "At its simplest, the event is combinatorial. Simplicity itself is very problematic: to grasp my statement as a 'combination', I have to disregard both the intentional movement that brought me to it and the exteriority in which it takes place. [...] When Saussure characterizes speech as a game of 'combinations by which the speaking subject uses the code of language', one must therefore see in it a theoretical fiction that only manages to pose the speaking subject as a 'user' of language by momentarily setting aside all intentional and enunciative dimensions of speech. I never, strictly speaking, 'use' combinatorial rules to arrange units drawn like dominoes or lottery balls from the linguistic 'treasure'. My speech first had to be a global aim before it could be analyzed in elements. It is only when I reconsider my 'realized', 'externalized' utterance that I can take an objective and detached look at it, and pretend that my discursive activity has been that of an assemblage." (Jenny 1990, p. 16, author's translation).

> particular meaning that must be sought. The linguistic signs of Kafka's *Metamorphosis* represent, through their signified, a certain state of affairs which, in turn, calls for interpretation. (Coseriu 1992, p. 76, author's translation)

More recently, in a semiotic tradition of Hjelmslevian descent, yet having broken with the algebraism of language theory, Rastier (2003, 2015) proposed a textual model of the sign in which meaning, considered essentially as the "network of relations between the signified within the text" (Rastier 2015, p. 89), seems to be able to be described in full in the usual terms of semiotic theory (bifaciality of units, differentiality, degrees of systematicity), reworked in a hermeneutic perspective (principle of contextuality, units considered as correlates of an interpretative activity) and reformulated in a gestaltist-inspired modeling framework (semiosis envisaged as a pairing of backgrounds and forms on both language levels)[4]. This "semiotic resolution" of the description of textual meaning, however, implies the secondarization of certain traditional questions, concerning, for example, language transitivity and its theticity (which Coseriu still considered in the classical terms of the philosophy of language), the defended pantextualism implying the relocation of the formulation of these problems in texts and intertext: "[...] The mysterious exterior of language is only the other side of the logico-grammatic restriction of linguistics: it resides in the texts and the intertext that it cannot conceive." (Rastier 2003, p. 42, author's translation).

The following discussion focuses more particularly on two aspects of this question of the semiotic articulation of textual meaning: first, and on the basis of the Coserian tripartition introduced above, we propose to distinguish three semiotic relations that constitute textual semiosis (*signification/designation/expression*), insisting on the need to distinguish for each of them the types of units that they relate on the two levels of language. In a second step, we will discuss the value of the notion of *significance* by detailing the ways in which these three relations are established in interpretative paths.

5.1. The articulation of meaning according to three semiotic relations: signification, designation and expression

5.1.1. *The relation of signification*

In a broad sense, the relation of signification implies the following properties for semiotic units:

---

4 Rastier gradually developed these proposals, a recent summary of which can be found in Chapters 3 and 4 of Rastier (2015).

1) they are *double* (inseparable union of a signifier and a signified);

2) *articulated threefold* (at the phonological, morphological and syntactic levels), each of these three levels define a semiotic type and admit units composed in *complexes*, the complexes of one level providing units for the higher level of integration[5];

3) their value is relational (all of these relations and the relations between relations forming the system);

4) and they are *differential*, the fundamental relations within the system being relations of difference.

In a more restricted sense, we have proposed (Missire 2018b, 2018c) to limit this relation of signification to the units of the lexico-constructional level, where their signified are most clearly attested, and to describe the semiotic relation of the morpho-phonological level as a *relation of significance* characterized by the indeterminacy of the signified of this level, and consequently expressing only the "signitive modalization" of the units, which are therefore already more than formants[6] and not yet signifiers. In the following, we will thus call the *relation of significance* the relation implying only the *recognition* of a unit as belonging to the linguistic system, and we will call the signification relation the *understanding* of the unit, that is, the concomitant realization of its signifier and its signified (see *infra* 2), which we have proposed to represent schematically as follows (Missire 2018c) (Figure 5.1).

The relation of significance is situated at the semiotic level of the Benvenistian division between semiotic and semantic modes and the relation of signification is at the semantic level. It is essential that this relation of signification does not imply an exit from language toward something that is designated and intended as these are

5 We follow here Bouquet (2013), whose work we will interpret enthusiastically for a detailed presentation of this triple articulation inspired by his interpretation of the Saussurean manuscripts: "The finality of the internal composition of one semiotic type can be seen as forming the minimal unit of another semiotic type. In other words: a plexus, produced by internal composition within a semiotic functional type, can be considered to be a maximum when its result coincides with the minimal unit of a sign of a distinct functional type. Thus, the purpose of phonemic composition (non-generative composition, i.e. composition that is fixed in a state of language and cannot be recursive) is to constitute the minimal unit of a morphemic sign. In turn, the purpose of morphemic composition (also non-generative, i.e. fixed, and non-recursive) is to constitute the minimal unit of a syntactic sign. Finally, the syntactic composition (generative and recursive) necessarily comes back, through the play of its functional principle called *fusion* in Chomsky's terminology, to the constitution of a maximum unit." (Bouquet 2013, p. 89, author's translation).

6 We use the term "formant" to refer to an acoustic/graphic concretization of a signifier, which is an abstract unit.

dimensions Benveniste reduces the semantic level of[7]. In this restricted sense, we will consider the value of a unit as the set of the signified (for example, the various acceptances of a lexical unit) to which so many relations of signification lead to, the signified network associated with a unit being specific to the semantics of each language.

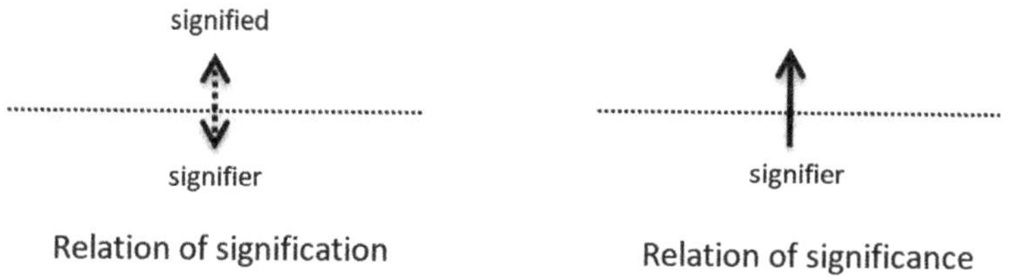

Figure 5.1. *Signification and significance*

5.1.2. *Designation*

The relation of designation, whose functioning is described by referential semantics, allows speakers to refer to objects of discourse, generically conceived as *themes* (independently of their quality as entities, processes or states of things), accessed because of the fundamental property of linguistic transitivity: it is impossible to speak without talking about something. At this level, the isolable linguistic fragments designating these themes of discourse should no longer be considered as *signifiers* engaged in a relation of signification, but as "designators" for themes outside the semiotic sphere. The definition of the sign as *aliquid stat pro aliquo* is appropriate for these designators, whereas it is inappropriate for *forms of meaning* at the signification level. If the regime of signification is observed at the lexico-constructional level, the regime of designation implies a minimal statement within which a "current" reference can be established, since the designators must be provided with a syntactic function[8]. The questions that usually arise at this level are

7 For a discussion, see Missire (2018c).

8 The thematic stratum that is the correlate of this semiotic regime only corresponds to a part of what Benveniste called the semantic mode, because he also included that which is "intended" in this mode, which we will discuss in the expressive regime below: "The semantic expression *par excellence* is the sentence. We say: the sentence in general, without even distinguishing the proposition, to stick to the essential, the production of the discourse. This time, it is no longer a question of the signified of the sign, but of what is referred to as the intended, that which the speaker wants to say, the linguistic realization of his/her thought. [...] With the sentence, one is connected to things outside the language; and while the sign has as its constituent part the signified that is inherent in it, the meaning of the sentence implies reference to the discourse situation and the speaker's attitude." (Benveniste 1974, pp. 222–225, author's translation).

those of *naming* entities and processes, of *identifying* them, and, in terms of statements, of determining the truth value of the propositions they contain.

5.1.3. *Relation of expression*

For lack of a better term, we will refer to what many authors call the *meaning* as the *expression*[9], noting that the interpretative paths do not end with the regimes of signification and designation but continue on this basis toward another semantic level.

Riffaterre (1983) had formulated it laconically but clearly with regard to poetic language – "a poem tells us one thing and means another" (p. 11, author's translation) – and this observation can be extended to all language productions: the derivation of a meaning from the designated does not only concern academic hermeneutics (plural meanings of writing, symbols of literature, etc.), and in the semantics of "ordinary" discourses, theoretical differences are marked precisely at the point of that which will be considered fundamental concerning the nature of this second scene on which we interpret the "real" meaning of what is being stated: is it identified with the argumentative value of the discourse? To the illocutionary and perlocutionary values? Or, on the contrary, only in terms of designated (referential semantics)? Will the symbolic have an economic basis (materialistic hermeneutics)? Pulsive (psychoanalytical hermeneutics)? Etc. We agree that these choices must be established within material hermeneutics (Rastier 2005)[10], and need to be arbitrated by taking into account the diversity of genres and discourses to which the texts in question belong (*critical* theme), the semioinguistic tools of such a hermeneutics simply having to provide the descriptive concepts to articulate the analyses (*descriptive/empirical* theme). On this last point, we proposed (Missire 2018b) to consider that, on both language levels, the correlates of the relation of expression could be described with the concepts of semantic and expressive backgrounds and forms (Rastier 2015), this choice allowing us to reinvest at this level in the descriptive concepts of the relation of signification (notably the concept of seme) to describe units (notably semic molecules) that are no longer linked to their correlate on the signifier plane by a reciprocal presupposition relation. Conversely, that which

9 Reserving, for our part, a broader extension to the latter term (see below).

10 Characterized by the three themes *critical*, *descriptive/empirical*, and *agnostic*: "The expression *material hermeneutics*, taken from Schleiermacher, denotes a full and ambitious form of critical hermeneutics in the philological tradition. This somewhat paradoxical name is justified in particular because this unification engages a reflection on the unity of the two planes of language, content and expression. Three main themes can be identified: the anti-dogmatic or critical theme; the anti-transcendental or descriptive (empirical) theme; and the anti-ontological or agnostic theme. They respond to the needs of a semantics that must consider the diversity of texts, within a semiotics of cultures [...]" (Rastier 2005, author's translation).

is expressed can be converted into the signified when it enters into a relation of signification associated with a semiolinguistic unit[11].

Thus, one should always distinguish in a given text in what way an *extract*[12] can be understood as the *signifier, designator* or *expressor*. For example, in the short text proposed *infra* (see section 5.2.2), the *rectangular* extract can be analyzed as:

1) the *rectangle* signifier, with realization in the context of the meaning containing the semes/geometric shape/, /right angles/, /4 sides/;

2) a part of the designation *a raw rectangle*, the designation of which remains undetermined at that point in the text;

3) a part of an expressor whose expression is the semic molecule {/volume/, /container/, /rectangular/, /liquid/, /blue/, /chlorinated/} which the text expresses discontinuously before the expressive summation "pool" at the end of the text. For a detailed summary, see Table 5.1.

	Types of semiosis	**Language levels**	**Correlates**	**Types of units**[13]
Semiotic articulation of the passage (textual semiosis)	Expression	Text	Expressed	Semantic backgrounds and forms (semic molecules, isotopes)
			Expressing	Backgrounds and expressive forms (phaemic molecules, isophonies)
	Designation	Statement	Designated	Themes (entities, processes, states, etc.)
			Designating	Syntagms with a syntactic case; statements
	Signification	Lexical constructions	Signified	Lexical signified, constructional, syntactic
			Signifier	Lexical forms, constructions, syntactic positions

Table 5.1. *Passage and types of semiosis*

11 The description of such conversions between signifier and expressive levels remains to be fully elaborated within the semantics of texts. For example, how can units such as specific semes (which, as the extremes of relations of opposition between semes within lexical classes) express the *local* character of any form of differentiality, find themselves identical to themselves in *translocal* units such as isotopes or semic molecules without being hypostasized and thus lose their yet defining character of expression of difference? For a discussion on these questions, see Missire (2005, pp. 147–164).

12 We use this generic term proposed by Rastier to designate the signifier of a *passage* (textual sign); its signified being the *fragment* (Rastier 2005, p. 81).

13 Given as an example, no thorough explanation here.

Since texts are composed of statements and these are lexico-constructional and syntactic units, the textual level includes relations of signification, designation and expression. When a statement is said to correspond to the level of the relation of designation, it is only *as a statement*. As a concrete statement, it is always a fragment of text, or a short text, and in this sense also includes relations of signification and expression.

5.2. Significance and meaning

Until now, we have proceeded as if these three types of semiosis were established "mechanically", through the application of rules internalized by the speakers: the relations of signification, designation and expression would thus be realized as a whole as soon as the interpreter had the linguistic, communicative and hermeneutic skills required for understanding the text. However, without even mentioning the accidents of productive or interpretative pathways[14], everyone has experienced the "resistance" of meaning, which is equivocated, delayed or, on the contrary, duplicated, and which is rarely realized throughout a monotonous course of action. The notion of *significance* has been one of the means that the disciplines of the text have used to objectify this property of "differance" in the realization of meaning. Let us briefly recall[15] that this notion was introduced and variously developed (Kristeva 1969; Meschonnic 1973; Barthes 1974; Meschonnic 1982, 1997; Lacan 1999) as a reaction, in particular, to a certain state of "structuralism"[16] in linguistics and in the disciplines of the text, which was considered incapable of restoring the "work of the text"[17] and, more broadly, all the phenomena pertaining to the subject's activity in language, an activity that introduces play into the regulated relations between the units of the system. Generally speaking, the relation of significance becomes perceptible through the "slowing down" of the interpretative pathways caused by their increasing complexity. With regard

14 Let us recall Culioli's (1990) famous formula according to which "understanding is a special case of misunderstanding" (author's translation).

15 We have detailed this analysis in Missire (2018c).

16 Moreover, it was largely fantasized by the "post-structuralist" authors of the 1960s (Rastier 2017).

17 "The text 'works', at every moment and from whichever side it is taken; even when written (fixed), it does not stop working, maintaining a production process. What does the text work on? Language. It deconstructs the language of communication, representation or expression (where the subject, individual or collective may be under the illusion that he/she imitates or expresses him/herself) and reconstructs another language, voluminous, without a background or surface, because its space is not that of the figure, of the picture, of the frame, but the stereographic one, of the combinatorial game, infinite as soon as one goes beyond the limits of current communication (subject to opinion, to doxa) and narrative or discursive verisimilitude." (Barthes 1974, p. 4, author's translation).

more particularly to the bifaciality of semiolinguistic units, the notion of significance implies the possibility of local breaks in the link[18] between language planes, and therefore invites us to describe phenomena that can be grouped together under the heading of problems of *aspectualization* of interpretative pathways. We have begun to refer to the relations between signification and significance *stricto sensu* as *supra* (see section 5.1.1). In what follows, we extend the examination of the relations between these two particular ones (section 5.2.1) and then propose to extend the notion of significance beyond its relative value to the relation of signification, in order to observe it in terms of relations of designation and expression.

5.2.1. *Significance and signification*

Within the framework of the semantics of texts, Rastier formulated a critique of the "monadic" model of the sign elaborated in the *CLGist* tradition, arguing the need to give precedence to a relational principle of intersign contextuality over the reciprocal signifier/signified intrasign presupposition relation usually considered characteristic of the Saussurian sign (Rastier 2015). He thus suggests substituting for the apocryphal monad of the CLG, a schema containing at least two related signs (Figure 5.2).

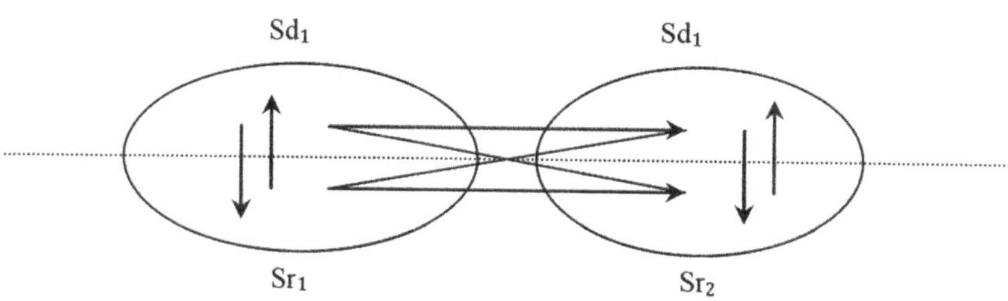

Figure 5.2. *The basic enunciative and interpretative pathways (Rastier 2015, p. 89)*

18 One of the first occurrences of which is undoubtedly to be found in the text "*L'instance de la lettre dans l'inconscient ou la raison depuis Freud*", published in 1957 in *La psychanalyse* before being taken up again in 1966 in Les *Écrits*, in which Lacan introduces this shift in the form of an *anticipation of the signifier* over the signified: "This is what will make possible an exact study of the links proper to the signifier and of the extent of their function in the genesis of the signified. [...] Only the correlations between the signifier and the signified provide the standard for any search for signification. [...] For the signifier of its nature always anticipates meaning by, as it were, unfolding its dimension before it." (Lacan 1999, pp. 494–499, author's translation). The term *significance* appears only twice in this text, once in a very generic sense: "this structure of language that makes possible the interpretation operation is at the principle of the *significance of the dream*, of the *Traumdeutung*" (Lacan 1999, p. 507, author's translation), and once in a technical sense: "S' designating in the context the productive term of the signifying effect (or significance) [...]." (Lacan 1999, p. 513, author's translation).

This diagram allows us to describe, on the one hand, intersign homoplane relations (sé1 <-> sé2 ; sa1 <-> sa2) and, on the other hand, intersign heteroplane relations (sa1 <-> sé2 ; sa2 <-> sé1) and intrasign (sa1 <-> Sé1 ; sa2 <-> sé2) relations, the former determining the latter (principle of contextuality):

> Two types of contextuality are established within the same plane (we can say they are homoplane). The pathway sé1→Sé2 recognizes a difference, or establishes either an elementary isotopy or an afference by seme propagation. [...] The sa1→sa2 pathway allows a contextual phonetic modification (e.g. liaison). [...] The sa1→sé2 pathway disambiguates one signified by the neighboring signifier. The converse path sé1→sa2 assigns a signification to the neighboring signifier, for example in the case of a rhyme. Priming provides many examples of these pathways: priming the sound of one item by the meaning of another, or vice versa, or even reciprocal priming. [...] As meaning consists essentially of a network of relations between signifiers within the text, signifiers can be considered as *interpreters* that enable the construction of some of these relations. (Rastier 2015, pp. 88–89, author's translation)

The elementary interpretative operations mentioned above allow for the elaboration of the concept of interpretative pathways at the textual level. However, the *interpreter* status conferred on signifiers, notably by the dissymmetry it creates between planes, evokes the problem of significance, Rastier did not explicitly develop his propositions in this direction. Bringing the notion of significance into play in such a framework consists of recognizing that, from the moment one substitutes these elementary interpretative operations for the reciprocal presuppositional signifier/ signified relation, it is necessary to take into account a possible aspectual shift between units of language planes: any signifier *recognized* in a path does not offer immediate access to a signified, but can simply *potentiate* it, or introduce the expectation of it without the latter being determined. From this point of view, whereas the relation of signification refers to the immediate signifier/signified conversion (i.e. a relation of reciprocal presupposition *form-meaning*), a relation in which the mode of existence of the units of the two planes is identical (the realization of one implies the realization of the other), the relation of significance recognizes differences between planes: starting from the signifier plane, the relation of significance corresponds, schematically, to the arrow[19] that starts from the signifier, an arrow that modalizes this signifier as a "signifier" without the necessary realization of a converse on the other plane (the realization of this signified, if and when it occurs, then "vanishes" the relation of significance). *Potential* and *indeterminate* are here the two signified characteristics in

19 The arrow, as opposed to the "bar" between signifier and signified, has been emphasized (Lacan 1999; Agamben 1998), because this bar blocking access does not allow us to account for the essential, which is the *modalization of a formant* as "signifier".

the relation of significance. We have proposed to consider that, with regard to linguistic units, we must distinguish at least three types of significance, depending on whether we envisage it (Missire 2018c):

i) At the *system*[20] level (*abstract virtual* level): the relation of significance (significance 1) points to the potential and indeterminate signified of the morphophonological units. This is the significance evoked by Benveniste when he raises the question of the *existence* of units at the semiotic level (versus the semantic level) (Benveniste 1966, 1974): a unit may or may not be *recognized* as existing, and therefore as signifying, independently of the *knowledge* of its signified[21].

ii) In terms of *norms* (current abstract level): the relation of significance (significance 2) corresponds here to the signified network associated with the lexico-constructional units (generalized polysemy), each of which, through the signifier, which then plays an interface role, can *evoke* the others[22].

iii) In terms of concrete textual productions (*concrete current* level): the relation of significance (significance 3) is equivalent to the homoplanar syntagmatic relations in terms of the signifier presented above. This type of signifier, which has been meticulously described by Meschonnic (1982), determines the two previous types (for example, a rhyme (significance 3) may elicit a syllepsis on one of the words in the rhyme (significance 2)).

Significance 1 is the fundamental relation, which remains present at all levels, but which the realization of the signified in terms of norms and texts (relation of signification) pushes into the background. Figure 5.1 can be completed schematically in this way.

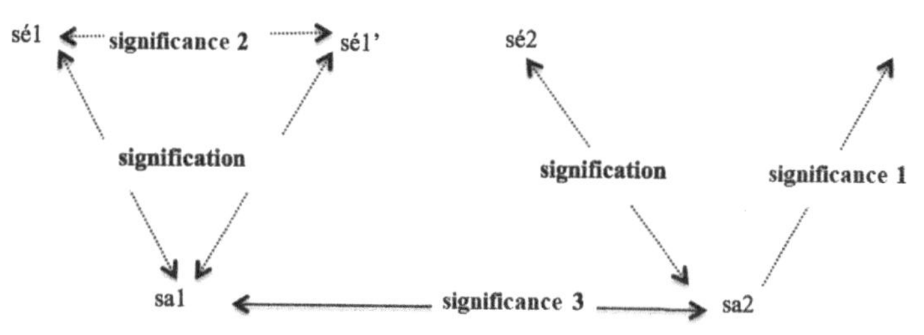

Figure 5.3. *Three types of significance*

20 Concepts understood in the sense of Coseriu (1952). For a more detailed presentation, see Missire (2004, 2018c).

21 For a discussion, see Missire (2018b, 2018c).

22 See, for example, this note by Barthes: "Significance [...] is indistinct at all levels of the work [...] in monemes, which are less semantic units than association trees and are carried by connotation, latent polysemy, into a generalized metonymy." (Barthes 1974, p. 7, author's translation).

NOTE.– The relation of significance 2 between sé1 and sé1′ is established through sa1, and thus presupposes significance 1.

And taking into account the three levels *system*, *norms*, and *speech*, we get Table 5.2.

	Abstract (type)		**Concrete (occurrences)**
	Virtual	Current	
Instances	System	Standards	Texts
Semiotic regimes	Significance 1	Significance 1 and 2 Signification	Significance 1, 2 and 3 Signification Expression Designation

Table 5.2. *Three types of significance and the three-level divide between system/norms/texts*

5.2.2. *Significance and designation*

If the relation of significance can be understood as what remains of the relation of signification once one has "subtracted" the signified from it, in the relation of designation it would then correspond to the relation of designation without the designated. In this case, it is access to all or part of the "scene" that is blocked, the themes escaping from a plenary perception even though they are presupposed by the thematico-predicative structures of discourse. This can be clarified by bringing back to this level of analysis the significance 1/significance 2 distinction introduced above: in the first case, an excerpt designates a theme without it being (immediately) accessible; in the second, it potentially gives access to several themes:

1) *Significance 1*: without claiming to be exhaustive, an example of this first type would be all phenomena involving variations of the naming routines used to designate a theme or process, in particular cases of:

- *semic deletion*, integral (see the analysis of things in Kleiber (1987)) or even pronouns: *that's all they think about*) or partial, particularly present in journalistic euphemisms (for example: "intervention" → "war"; "renewal" → "expulsion"; "operations" à "bombings", etc.) (Bonhomme 2005);

- *isotopic transposition*, especially in metaphorical connections ("sickle" → "Moon");

- *denominative expansion* (pronomination, periphrasis), as in this short text by Chevillard, where the lexicalization of “pool” is delayed while its semic constituents are progressively lexicalized from the beginning[23]:

> One must imagine the gentle Tuscan hill being stupidly/[pejorative]/ broken[/dysphoric/] by a rectangle[/rectangular/] raw [/dysphoric/] – heresy [/dysphoric/ /pejorative/] geometric/[geometric shape/].
>
> We must imagine this [pejorative]/garish[/dysphoric/ /pejorative/] touch of blue[/blue/] lagoon[/liquid/], of blue[/blue/] chlorinated/[chlorinated] on the side of the Tuscan hillside where the silver green of the olive tree and the black of the cypress blend so well, in the golden light, against the backdrop of Siena's land – heresy [/dysphoric/ /pejorative/] chromatic[/blue/].
>
> You have to imagine this/[pejorative] absurd/[pejorative/] pool/ [rectangular/ /blue/ /liquid/ /chlorinated/]. And me in it. (Éric Chevillard, *autofictif*, author’s translation)

These elementary processes can be combined, as in the Scandinavian *kenningar* (expansion and transposition), or in this example of the surrealist game *L’un dans l’autre* (meaning: One in the other) where textualization mobilizes transposition and expansion to enigmatize the theme to be found (here *burrow*):

> I’m a flowerpot average in diameter. The plant I am sheltering has rejected all my soil outside so that it can find its comfort and activate the circulation of its sap from the inside to the outside and vice versa[24].

2) *Significance 2*: in addition to the usual cases of ambiguity in the resolution of anaphores, one can evoke more complex operations in which the referential impression becomes opaque due to the potential multiplication of designations. For example, in Baudelaire’s quatrain, *Tristesse de la lune (Sadness of the Moon)*:

> Tonight the moon dreams with more indolence,
> Like a lovely woman on a bed of cushions,
> Who fondles with a light and listless hand,
> The contour of her breasts before falling asleep;
>
> On the satiny back of the billowing clouds,
> Languishing, *she* lets herself fall into long swoons,
> And casts her eyes over the white phantoms
> That rise in the azure like blossoming flowers.

---

23 For a more detailed analysis, see Missire (2018b).

24 For a detailed analysis of the mechanisms involved, see Grea (2010).

> When, in her lazy listlessness,
> She sometimes sheds a furtive tear upon *this globe*,
> A pious poet, enemy of sleep;
>
> In the hollow of his hand catches this pale tear,
> With the iridescent reflections of opal,
> And hides it in his heart afar from the sun's eyes.
> (translation by William Aggeler; italics added by the author.)

The antecedent of "she" in verse 6 is syntactically "the moon" in verse 1, but the syntactical complexity of the period, the distance between anaphorizing and anaphorized, the isotopy between the processes associated with "she" ("swoons", etc.) and "lovely", and the comparative/metaphorical context between "the moon" and "a lovely woman" can induce an interpretation in which the latter, syntagmatically closer, substitutes itself for "the moon". Comparably, "this globe" in verse 10 certainly refers to the earth (deictic value, place of indexing of the enunciative focus "a poet") but also evokes (anaphoric value) the "breasts" of verse 4, regularly referred to as such in Romantic poetry[25].

5.2.3. *Significance and expression*

The question of significance in terms of the relation of expression as we have defined it is at the very basis of the hermeneutic question, especially in our cultural areas where an expressive tradition has taken hermeneuticity and interpretative difficulty as a principle (Agamben 1998). However, care must be taken not to confuse this "differance" in the establishment of the relation of expression with the one encountered in terms of the linguistic sign in the relation of signification, because here it is obviously the problem of the symbol and its opacity that is at stake, as already seen in connection with the meaning of *The Metamorphosis* in Coseriu's distinction given in the introduction. Without confusing the semiotic and symbolic levels, Lyotard clarified the difference between the two, denying the *phenomenological* quality of the sign to the units of the former and reserving it for those of the latter:

> [...] the designated thing is not a thing but a symbol, and it is legitimate to say at the outset that it is opaque. [...] Opacity is in the object, not in the word, nor in its distance from the object. Words are not signs, but as

25 For example, Lamartine: "From her virginal breast she covered the globe" (*La chute d'un ange*, 15e vision); and at Gautier's, the dedicatee of *Les Fleurs du Mal: "Son sein, neige moulée en globe*" (meaning: Her breast, snow moulded into a glove) (*Emaux et Camées*, "*Symphonie en blanc majeur*"), "*faisant jaillir ta gorge en globe*" (meaning: "Making your throat gush like a globe") (*Emaux et Camées*, "*Une robe rose*"). For a detailed discussion of these phenomena of referential ambivalence in this poem see Missire (2005).

> soon as there is a word, *the designated object becomes a sign*: that a designated object becomes a sign means precisely that it conceals a "content" hidden in its manifest identity, that it reserves another face for another view of it. Words or linguistic units are not signs by signification; they are not signs by designation; but they make signs with the objects they designate (make see) and signify (make hear), and from which they are separated. (Lyotard 1971, p. 82)

For our part, we have seen that what we have called significance 1 and 2 showed a comparable potential opacity in terms of the relation of signification, the dissipation, or even the maintenance of the latter being governed by taking into account the general-discursive nature of the texts.

In conclusion, we will propose a definition of *textual meaning*, constituting (1) the correlates, in terms of the signified, of the relations of signification designation and expression; and (2) the relations between these correlates[26]. We will also define *textual semiosis* as the set of relations of signification, designation and expression, as well as, for each of them, the possibility of a relation of significance.

Given the complexity of the relations and the units involved, it seems pointless to retain, if not for didactic purposes, the notion of a "textual sign": the *passage* is thus primarily a plan for integrating semiotic units, but is not itself a sign.

The necessary conversion relation between units of the relation of expression and of the relation of signification through those of the relation of designation is a *constitutive* limitation for a semantics of texts with semio-linguistic foundations: that which cannot enter into this cycle of conversion from the expressive level remains inaccessible to such a semantics, and must be taken care of at another level. This is an opportunity to recall that access to meaning is at best asymptotic, and that no discipline can exhaust its description. Material hermeneutics could be the place where the diversity of disciplinary approaches could be articulated in a "reasoned eclecticism" (Gérard 2018).

With regard to the Benevolent division between the semiotic and semantic levels that we mentioned in the introduction, we must therefore conclude that such a conception, with its requirement of the "two linguistics", has concluded a bit hastily from the impossibility of saying *everything* from the textual meaning in semiotic terms to the impossibility of saying *nothing*, thus depriving itself of access to the various modalities according to which semiotics exists in speech[27].

---

26 The latter we have not mentioned here. For a description of the designated and expressed relations on the model of the relations between *theme* and *scheme*, see Missire (2018b, pp. 3–4).

27 For further arguments in support of this conclusion, see Missire (2018c, pp. 4–5).

NOTE.– This work belongs to the same research cycle on significance as two other studies – Missire (2018b) and Missire (2018c) – which the reader may refer to for further study.

5.3. References

Agamben, G. (1998). *Stanze*. Rivages, Paris.

Barthes, R. (1974). Théorie du texte. In *Encyclopædia Universalis* [Online]. Available at: https://www-universalis-edu-com.nomade.univ-tlse2.fr/encyclopedie/theorie-du-texte/ [Accessed 31 August 2019].

Benveniste, É. (1966). *Problèmes de linguistique générale 1*. Gallimard, Paris.

Benveniste, É. (1974). *Problèmes de linguistique générale 2*. Gallimard, Paris.

Bonhomme, M. (2005). *Pragmatique des figures du discours*. Champion, Paris.

Bouquet, S. (2013). Triple articulation de la langue et articulation herméneutique du langage. In *De l'essence double du langage et le renouveau du saussurisme*, Rastier, F. (ed.). Lambert-Lucas, Limoges, 81–93.

Cadiot, P., Visetti, Y.-M. (2001). *Pour une théorie des formes sémantiques. Motifs, Profils, Thèmes*. PUF, Paris.

Coseriu, E. (ed.) (1973). Sistema, norma y habla. In *Teoría del lenguaje y lingüística general, cinco estudios*. Gredos, Madrid.

Coseriu, E. (1997). *Linguistica del testo. Introduzione a una ermeneutica del senso*. Carocci, Rome.

Gérard, C. (2018). Art du langage et linguistique du sens. Traduction commentée des "thèses sur le thème 'langage et poésie'" (Eugenio Coseriu). *Pratiques*, 179/180.

Gréa, P. (2010). Je suis un pot de fleur de diamètre moyen : énigme et perception sémantique. *Texto ! Textes et Cultures. Revue de l'Institut Ferdinand de Saussure*, xv(3), 1–18.

Jenny, L. (1990). *La parole singulière*. Belin, Paris.

Kleiber, G. (1987). Mais à quoi sert donc le mot chose ? Une situation paradoxale. *Langue française*, 73, 109–128.

Kristeva, J. (1969). *Sèméioitikè. Recherches pour une sémanalyse*. Le Seuil, Paris.

Lacan, J. (ed.) (1999). L'instance de la lettre dans l'inconscient ou la raison depuis Freud. In *Écrits 1*. Le Seuil, Paris, 490–526.

Lyotard, J.-F. (1971). *Discours, figure*. Klincksieck, Paris.

Merleau-Ponty, M. (1969). *La prose du monde*. Gallimard, Paris.

Meschonnic, H. (1973). *Pour la poétique III. Une parole écriture*. Gallimard, Paris.

Meschonnic, H. (1982). *Critique du rythme. Antropologie historique du langage*. Verdier, Lagrasse.

Meschonnic, H. (1997). Benveniste: sémantique sans sémiotique. *Linx*, 9, 307–326.

Missire, R. (2005a). Sémantique des textes et modèle morphosémantique de l'interprétation. PhD Thesis, Université de Toulouse II [Online]. Available at: http://www.revue-texto.net/Inedits/Missire/Missire_These.html [Accessed 28 March 2018].

Missire, R. (2005b). Une larme baudelairienne, essai de description morphosémantique de Tristesses de la lune. *Champs du signe, concours et recherche, CAPES et Agrégation de Lettres*, 20, 87–114.

Missire, R. (2018a). Unités linguistiques d'une sémantique discursive. *Langages*, 210(1), 17–34.

Missire, R. (2018b). Flux psychique, sémiosis langagière et niveaux de l'analyse linguistique. *Le symbole est-il diabolique ? Signifiant/ signifié: la duplicité du signe en question*, 2, 207–225.

Missire, R. (2018c). Faire sens et avoir un sens. Note sur la signifiance linguistique. *Pratiques*, 179–180.

Rastier, F. (2003). Le silence de Saussure ou l'ontologie refusée. In *Saussure*, Bouquet, S. (ed.). L'Herne, Paris, 23–51.

Rastier, F. (2005). Herméneutique et linguistique : dépasser la méconnaissance. *Texto ! Textes et cultures*, x(4) [Online]. Available at: http://www.revue-texto.net/Dialogues/Debat_Hermeneutique/Rastier_Herm-et-ling.html [Accessed 8 November 2018].

Rastier, F. (2015). *Saussure au futur*. Les Belles Lettres, Paris.

Rastier, F. (2017). Cassirer et la création du structuralisme. *Texto ! Textes et cultures*, xxii(4), 1–15 [Online]. Available at: http://www.revue-texto.net/index.php?id=3977 [Accessed 19 August 2019].

Riffaterre, M. (1983). *Sémiotique de la poésie*. Le Seuil, Paris.

6

Semiotics of Cultures and Theoretical Hybridities: For a Renewal of Thought

We live in the oblivion of our metamorphoses.

Paul Éluard
(Éluard 1946, p. 43, author's translation)

For the first time in the history of humankind, human societies have demonstrated plurality in their way of thinking and functioning: they are multicultural, multilingual, cosmopolitan, mobile and hybrid. Technological progress and abundance of media offers have changed the ways of thinking, working, informing, entertaining, buying and learning, and both the rhythm and pace of change have accelerated considerably. Today, in order to be precisely analyzed, the slightest societal phenomenon calls simultaneously for several specialists from various academic fields. Analyzing a dynamic and changing humanity with monodisciplinary theoretical tools, moreover from previous centuries, which in the time span of a century has become so diversified, inevitably leads to incomplete or incorrect analyses[1].

Rethinking theoretical and terminological tools in order to recompose them and adapt them to this constantly changing new society, taking into account the scientific progress of the last 30 years, is, for the human and social sciences (HSS), the great challenge of

---

Chapter written by Astrid GUILLAUME.

1 We think, for example, of philosophical works that deprive animals of thoughts, dreams, intelligences, cultures, intentions, languages and consciousness due to a lack of suitable inter-theoretical tools and advanced technologies, overly rigid definitions, ignorance of research carried out in the life sciences, or religious or species-dominated ideology. This thinking brings together semiotics, ethology and philosophy and makes it possible to question these archaic philosophical approaches.

the 21st Century. It is also a matter of scientific ethics to point the finger at the hermeneutical errors of the past, sometimes still well anchored today, and to question erroneous approaches in order to better rethink them with new theories and new terms.

This is one of the roles that the cultural sciences are taking on (Rastier and Bouquet 2002), as part of their knowledge of the cultural interactions of the human and non-human worlds[2], in order to analyze these multidisciplinary and multifactorial societal phenomena as accurately as possible.

Over the last 50 years, the fact that the academic disciplines in the HSS have become compartmentalized, with a predominance of mono-disciplinarity in academic curricula and careers, has deprived entire scientific fields of theoretical innovation. A long intradisciplinary approach has made the academic-based HSS methodologically inflexible and sometimes fruitless in their inability to analyze radical, interdisciplinary phenomena. The lack of flexibility in terminology and the lack of knowledge of the progress made in neighboring scientific subjects have kept many disciplines in the field of HSS out of the debate, sometimes even endangering them in university curricula due to a lack of accessibility and adaptability to current dynamics (Guillaume 2010).

Thus, being un-disciplined has become an emergency for scientific innovation in the HSS. For the cultural sciences to make sense and make science, they must dare to move beyond the rigid, fixed, monodisciplinary academic framework that characterized the last century. They introduce the beginnings of new approaches and establish new lines of research, neologisms and even new disciplines by emancipating themselves from established definitions and disciplines that are sometimes obsolete in view of the advances made in other fields of knowledge (medicine, physics, zoobiology). Indeed, thinking about new hybrid and transdisciplinary theories is a way of making complete scientific analyses of the facts of society, themselves multidisciplinary by definition, and at the same time a way of forging new words (Guillaume 2019a).

Opening up to multitheoricity, intertheoricity and transtheoricity means enabling a renewal of thinking and analytical prisms in the HSS. In the cultural sciences, François Rastier's theories lead to this crossroads of academic disciplines, enabling theories to be conceived as cultural objects with plural scientific origins and a wide range of promising futures. For an overview of the rich fields of action of François Rastier's theories, refer to Ablali *et al.* (2014).

---

2 By non-humans, we mean living beings in their great diversity (animals and plants) as well as artificial intelligences (humanoid and animaloid robots) or present and future hybrids (transanimalism).

6.1. Theories: cultural objects in transfer

Theories, abstract thought processes, are projected and constructed in synchrony as well as in diachrony. They make *sense* and *science* at the same time. Three types of theories emerge from the fields of knowledge: cumulative theories (number theory), static theories that evolve only because of their author (A.J. Greimas) and hybrid theories that metamorphose in contact with other disciplines (F. Rastier). This theoretical diversity is a form of richness for the language sciences, but it is above all the third range of theories that enables the most diverse and significant advances.

While intertextuality, intermediality and interartiality are terms and practices of analysis that have long been recognized, intertheoricity or hybridization of theories remain seldom practiced in the HSS in general and in the language sciences in particular. It nevertheless paves the way for new ways of thinking about societal phenomena and allows for the creation of innovative multidisciplinary, interdisciplinary and transdisciplinary theories. In order to assess intertheoricity, three characteristics must be brought together: first, the elasticity of the source theory/theories subject to theoretical variations in time and space; second, theoretical plasticity, which goes hand in hand with geometric modeling; and finally, theoretical hybridity, the final but never definitive result of inter, multi- or transtheoretical fusion. For a more detailed development on elasticity, plasticity and hybridity, the reader can refer to Guillaume (2014a).

While cultural objects are transformed according to the contexts that surround them (Rastier 2011b; William 2015b), theories in the HSS have so far been little recomposed, they are not very elastic. Elasticity here defines a theoretical and praxeological stretch, one that is inter-, multi- and transdisciplinary but also spatial and temporal. This implies a transformation of certain aspects of the theories and the partial maintenance of their terminology in the newly created theory. However, some theories do not transform, they seem to be fixed and devoid of elasticity. This may be because many theories are trademarked or partly patented. Their authors do not suffer the slightest change in their thinking. Anyone who would risk inverting or even mixing the terminology of two different or opposing schools of thought would run the risk of antagonizing the representatives of both schools. However, in sciences (medicine, zoobiology, physics, robotics, etc.) and also in arts (painting, music, fashion, sculpture, digital art, etc.), in diachrony as well as in synchrony, it is by mixing theories, media and genres that significant scientific and artistic advances have been achieved through contact with contemporary thinkers and/or those of previous centuries.

In language sciences, theories more rarely enter into a cycle of hybridization or recomposition, they have a slower evolutionary process, and are mostly autonomous and isolated. Researchers in the language sciences work little on the lab bench as do those in the life sciences. For example, they publish less as a team than researchers in biology or medicine. As a result, they create their own theories and terminology, which are rarely intended to change over time. On the contrary, the more static they remain, the more faithful to their "creator" they become. Moreover, for a long time they were only monodisciplinary, mainly textual fields. Only recently the language sciences have opened up to multidisciplinarity, interdisciplinarity and transdisciplinarity. This implies the acceptance of other disciplinary fields, other terminologies with other definitions, and terminological and cultural recontextualizations, which are indispensable to the transdisciplinary intertheoretical approach. This opens up new scientific horizons, and even new fields and disciplinary axes such as semiotraductology, traductogenesis (Guillaume 2017b), transferogenesis (Guillaume 2014a), biosemiotics, zoosemiotics (five research axes have been promoted by the French Society of Zoosemiotics (2019)), phytosemiotics, agrosemiotics[3] and humanimalism (Guillaume 2017c, 2019c).

However, where intuition plays a fundamental role in the arts (Kandinsky), philosophy (Bergson) and science (Newton, Schrödinger), it is unwelcome in the field of linguistics. Although, this open-mindedness in the arts and sciences gives theories their potential for metamorphosis and hybridity. HSS theories evolve under the impulse of a thinker but rarely via the osmosis of different theories, let alone in relation to feeling or perception. Although the phenomenology of perception has been witnessed with Merleau-Ponty's work, and although Peirce's phaneroscopy is now considered a phenomenology, the influence of art or geometry on HSS theories is not common, and these collusions are rarely made. In spite of this absence of colliding disciplines, the thoughts of certain theorists have experienced a fine renaissance after their authors' death, but mostly in a monodisciplinary manner: this is the case of Ferdinand de Saussure's theories, which witnessed a long progression of applications after his death until today – we now speak of neo-Saussureism – of Peirce's theories or even of Greimas' theories introduced into secondary school curricula, a fact that is rare enough to be highlighted.

François Rastier's theories open up many fields of knowledge. They demonstrate that they have a *plastic*, *elastic* and *hybrid* character, in the sense that they can be translated[4], transferred and made to evolve elsewhere and differently, but also because they are theories which, generating hybridity, carry within them this hybridity and this openness to otherness, necessary conditions for intertheoricity.

---

3 Agnès Alessandrin's doctoral thesis was defended at the end of 2019 at the Université Paris Descartes, it bears the title *Du gène au signe du gène. Recherche croisée pour la définition et le cadrage du champ de l'agrosémiologie* (meaning: From the gene to the sign of the gene. Cross-referenced search for the definition and framing of the agrosemiology field). It heralds an opening up of language sciences and semiotics toward genetics applied to the field of agronomy.

4 *Translated* to be understood here as the passage from state A to state B.

6.2. Definitional reminder

The cultural sciences are by definition *multidisciplinary*, *interdisciplinary* and *transdisciplinary* because they concern human and non-human cultures in a whole that involves their spheres of influence (William 2017, p. 7) as well as their history, their contexts and their future. The three terms *multidisciplinary*, *interdisciplinary* and *transdisciplinary* are, however, often confused or even misused. However, they are not synonyms. We shall retain the definitions given by Basarab Nicolescu, physicist, founder and now Honorary President of the International Center for Transdisciplinary Research (CIRET 2019).

Multidisciplinarity	**Interdisciplinarity**	**Transdisciplinarity**
Multidisciplinarity concerns the study of an object of one and the same discipline by several disciplines at the same time. For example, a painting by Giotto can be studied through the lens of art history crossed with that of physics, chemistry and the history of religions, European history and geometry. Or, Marxist philosophy can be studied through the lens of philosophy, which crosses with physics, economics, psychoanalysis or literature. The object will thus emerge enriched by the hybridization of several disciplines. The knowledge of the object in its own discipline is deepened by a fruitful multidisciplinary contribution. Multidisciplinary research brings a *bonus* to the discipline in question (art history or philosophy, in our examples), but this "bonus" is at the exclusive service of that same discipline. In other words, the multidisciplinary approach goes beyond the disciplines, but *its purpose remains within the framework of disciplinary research.*	The ambition of *interdisciplinarity* differs from that of multidisciplinarity. *It concerns the transfer of methods from one discipline to another.* Three degrees of interdisciplinarity can be distinguished: (a) *a degree of application*. For example, the methods of nuclear physics transferred to medicine lead to the development of new cancer treatments; (b) *an epistemological degree*. For example, the transfer of the methods of formal logic into the field of law generates interesting analyses in the epistemology of law; (c) *a degree of engendering new disciplines*. For example, the transfer of mathematical methods into the field of physics has engendered mathematical physics, from particle physics to astrophysics – quantum cosmology, from mathematics to meteorological phenomena or those of the stock market – chaos theory, from computer science into art – computer art. Like multidisciplinarity, interdisciplinarity transcends disciplines but *its purpose also remains embedded in disciplinary research*. By its third degree, interdisciplinarity even contributes to the disciplinary big bang.	*Transdisciplinarity* concerns, as the prefix "trans" indicates, *what is both between disciplines, across disciplines and beyond any discipline*. Its purpose is *the understanding of the present world*, one of whose imperatives is the unity of knowledge.

Table 6.1. *Nicolescu's (1996) definitions of multidisciplinarity, interdisciplinarity and transdisciplinarity*

These three definitions, which are essential for a good understanding of what multidisciplinarity, interdisciplinarity and transdisciplinarity include, make it easier to grasp the meaning and the many scientific backgrounds implied by the words *multitheoricity*, *intertheoricity* and *transtheoricity,* which we define accordingly.

Multitheoricity	Intertheoricity	Transtheoricity
Multitheoricity refers to *the study of the same subject by several theories* from a single discipline or from several disciplines at the same time. *Its purpose remains within the framework of theoretical research on a single subject.* In this case, the angle of approach differs according to the disciplines and the theories used, but the subject studied remains the same. *Multitheoricity consists of studying the same subject but from several theoretical angles.*	*Intertheoricity concerns the transfer of theories from one discipline to another with or without modifications of objectives.* Human medicine and veterinary medicine use the same theoretical tools in many cases but with different results because there is a different type of patient (different heart rate, different blood formulas and feeding patterns, etc.). Another example is a semiotician who integrates the translator's theories to apply some of Peirce's or Rastier's theories to create a new theoretical model[5]. Or, currently, mathematics is being combined with chronobiology to model the effectiveness of cancer treatments without the use of *in vitro* or *in vivo* tests. *Intertheoricity allows the creation of new theories.*	*Transtheoricity is both between the theories, across the different theories and beyond any theory.* It generates a theoretical thought that goes beyond the disciplines to which it initially belongs and ontological universalism by an ethical universalism that often needs to be developed and integrated by all. It implies universal ethics, to which a maximum adherence is required. A golden rule such as *don't do unto others what you don't want done unto you* pertains to religious, philosophical, legal, sociological, transotological and historical transtheoricity, etc. *Humanimalism*, for example, is fully a matter of transtheoricity. It is an ethical humanism that respects otherness and biodiversity and is rooted in transdisciplinarity. *Transtheoricity opens up to an ethical universalism that respects the living.*

Table 6.2. *Definitions of the terms multitheoreticity, intertheoreticity and transtheoreticity*

These three approaches carry meaning and science for the cultural sciences, and more particularly for the semiotics of cultures. *Multitheoricity* offers enriching multidisciplinary encounters around a theme, all the more so as this kind of interscientific encounter has been lacking for several decades. *Intertheoricity* allows theoretical innovations through hybridization or innovative recomposition, which implies rethinking scientific definitions and approaches (zoosemiotics, semiotraductology). Finally, *transtheoricity*, which is deeply rooted in universal ethics (William 2015a),

5 For an example of an intertheoretical study where Peirce's semiotics is applied to ethology and generates a new theoretical model in zoosemiotics, refer to Delahaye (2019).

is oriented toward programs for the renewal of thought in a pacifying perspective (William 2011) such as humanimalism (2019c), which redefines a humanism that respects sentient and sensitive living beings and biodiversity as a whole.

For this chapter, we favor intertheoricity, which enables real epistemological, hermeneutical and theoretical innovations.

6.3. Status of the arts and religious sciences

Intertheoricity opens up the disciplines and blurs overly rigid disciplinary outlines, breaking down the boundaries of the fields of human knowledge by bringing together the humanities and the sciences, but also the arts and spiritualities of the world, long sidelined by scientific and theoretical approaches. Indeed, while the arts and spiritualities have not always been considered as disciplines in their own right, they are today recognized in universities as objects of "science" (for the sciences of art, see Colombat 2017); for the religious sciences (École Pratique des Hautes Études 2019), sciences understood here as the production of the human spirit to be studied historically. Religious sciences do not confer any scientific status to spiritualities, they allow their contextual process to be analyzed in time and context. As such, it is important to integrate them into societal analyses, of which they are a part of. The same applies to the art sciences. Art does not acquire any scientific status, but its history, its various practices and multiple formats are part of human thought and history. Art theories have made history, and Erwin Panofsky remains a worthy representative.

Having been for a long time rejected by the scientific and theoretical spheres, the influence of arts and spiritualities on cultures but also on the creation of theories is nevertheless major. This influence can take place at different levels:

– the first pre-disciplinary level puts the intuitive before the scientific; a particular literary, esthetic or sacred interpretation favors a particular intuition or a particular scientific or artistic work (Kandinsky);

– the second level favors only the scientific fact or puts it before the intuitive, it will study the history of the discipline or certain elements over time according to different contexts (Einstein)[6];

– the third level mixes the two previous levels alternately (Merleau-Ponty, Bergson, Peirce).

6 This approach in no way excludes the possibility that a theory that is scientifically accepted by all at one time may not be accepted again in the following century. For example, Newton's theories are currently being challenged by Andrea M. Ghez's team at UCLA, not because they are no longer relevant, but because they did not address phenomena such as black holes. This confirms the theory of transferogenesis (Guillaume 2014a).

In any case, as fields of human knowledge, the art sciences and religious sciences influence human thought and interact implicitly or explicitly in the theorization of HSS. They play a role in the semiotics of cultures. These interdisciplinary contacts linking arts, sciences, religion, human and social sciences, technology, human and non-human medicine, i.e. *human and non-human contexts*, open up to different approaches that make it possible to study more precisely the cultural sciences as described in the work of François Rastier (2002, 2011) in connection with a linguistic anthropology.

6.4. Geometric plasticity of theories

Even the most monodisciplinary theories do not avoid a modeling approach to geometry, and thus indirectly to mathematics and logic. Theoretical geometric modeling is sometimes more esthetic than purely mathematical, but its omnipresence in human thinking in all disciplines deserves a moment of attention.

Indeed, geometric modeling seems to be an essential part of theorization. To give rise to a theory requires a precise process that brings together a precise terminology to be applied to the letter and to the number, depending on the discipline, one will find mathematical and physical formulas, diagrams that model or summarize the theory in the form of a figure, which is often geometric or at least geometrical. Rare are the theories that are not accompanied by geometrizing. In linguistics, these geometric models visually summarize a narrative, discursive theory, a tension, a perspective and a process that develops over time, thus offering additional visual and spatial precision to a textual development. They contribute to giving theories their scientific character and giving them a semantic *plasticity*.

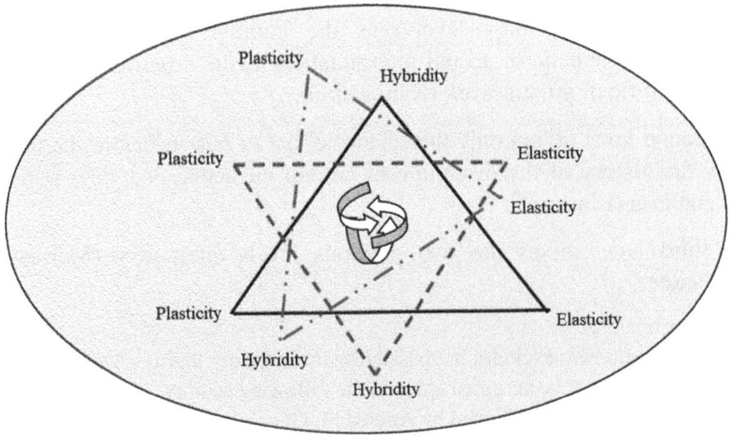

Figure 6.1. *Tri-triangular dynamics of intertheoricity with ethical background*

For example, here, the tri-triangular dynamics of intertheoricity presents the dynamic step, the core, of one of the processes in action within the framework of a theoretical transfer. To see all the stages at work in transferogenesis, see Guillaume (2014a). So that the transfer from one theory to another can take place, the three characteristics, elasticity, plasticity and hybridity must be able to co-exist at each moment of the transfer process. This implies a dynamic of composition/decomposition/recomposition bringing together *plasticity* in relation to the modeling of the first theory, *elasticity* in relation to the transfer of context (in diachrony and/or synchrony) to a new theory and *hybridity* in relation to the plasticity of the new theory acquired via this transfer. The theory then responds to the same process that composes, decomposes and recomposes cultural objects.

Geometry, a symbol of precision, is the reflection of what is present at the microscopic and macroscopic scales, at the scale of the visible and the invisible, the infinitely small and the infinitely large. In the fields of knowledge, everything is based on and relates to geometry and its forms (point, line, circle, semicircle, curve, triangle, rectangle, plane, depth, space, axis, level and dynamics).

On the spiritual and religious level, geometry has been widely practiced in different forms, from architecture to writing: the pyramids of Egypt, giant tombs allowing the one who rests there to reach the afterlife, remain internal and external geometric jewels; Buddhists and Hindus, via cosmogonic mandalas, put geometry forward; the Jewish and Arab cultures are attached to gematria, derived from the Greek word meaning geometry; medieval Christian cathedrals are showcases of geometrical calculations, strategic orientations in relation to space and light, rosettes and stained glass windows; the Muslim religion gave geometry supreme status as proof of a divine presence. Associations of numbers and letters, architectural buildings or signs, and works of art symbolize in all these religious contexts the proof of the existence of a creator and the perfection of His creations and creatures. For the cultural sciences, the question is not one of belief but one of knowledge of these worship-related, cultural, historical and artistic facts and signs and of knowing how to recognize them in their original or adapted forms in the different contexts where they may resurface today, whether or not they are decontextualized from the initial thought.

In science, the atom, matter, DNA, chromosome and space are represented by lines, axes, curves and connecting points, giving everything its own identity, a signature that distinguishes and creates the uniqueness of any matter, any theorem, any animate and inanimate being. Changing an atom, modifying the curve of a line or the place of a point by one micrometer, permutation of a figure, moving a comma, and everything then becomes other or no longer exists: a plurality of units that leads to a diversity of units, and vice versa to infinity. Abstract geometric representation is also part of scientific imagery, which promotes its popularization. When a CNRS team studied the phenomenon of cyclones using the evolution of soap bubbles subjected to different phenomena of instability, it recreated, in so doing, on a smaller scale, the turbulence of cyclones: this passage from the infinitely large to another scale, more accessible but perfectly comparable in terms of physical reactions, is an example of a transdisciplinary theoretical transfer of space, time and form that represents, on the elastic and translatological level, the perfect scientific transfer that makes sense and makes science. The scientific and artistic overcrowding of abstract geometric art, present on the microscopic and macroscopic scales and symbolically at the heart of everything and everyone, unites all academic fields. It is as if all the abstract artists of the early 20th Century had demonstrated an intuition of this invisible geometric form even before science could fully demonstrate it (Anourhy 2014; Centre Pompidou 2019).

Indeed, in the domain of arts, from ancient Egypt to Kandinsky, and even more recently, in inter-medial choreographies such as the Japanese group Enra (ENRA 2019)[7] and Miguel Chevalier (Chevalier 2017; Art Fractal 2019), scientific art creations were also geometry. Geometry has made its entry into contemporary choreographic art as well as into architecture, orientation, innovative designs and creations, perspective and depth of field, symbolism and color abstraction.

> The theory of fractals was first developed by the mathematician Benoît Mandelbrot and presented in 1975 in his seminal book: *Les Objets Fractals*.
>
> This mathematical theory, which sought to account for complex figures and objects, then spread to many disciplines including the plastic arts.
>
> Fractal art has thus become a powerful model of thought and a new visual code, just as cubism, abstraction, kinetic art and concrete art were. (Art Fractal 2019, author's translation)

Intermediality has made it possible to cross and interweave these spheres of application by multiplying the possibilities. Bridget Riley sets up implicit and explicit dynamics that bring together the play of contrasts, perspective and geometry, a process that is specific to kinetic art. Auguste Herbin elaborates a "plastic alphabet", associating letters, shapes, colors as well as musical notes. In *Danseuse* (1942), he develops the following alphabet: D (light red circle, do, re); A (pink, combination of circular, triangular, semicircular and square shapes, do, re, mi, fa, sol, la, si); N (white, combination of all shapes and notes like A); S (intense blue-green, combination of semicircular and triangular shapes, la, sol, fa); E (red, circle, do); U (blue, semicircle, sol, la). What the artist is looking for here is to

7 The Japanese group ENRA makes geometric multimedia and intermedia choreographies (Performing Arts Collective) that mix arts, dances, ballets, video games, etc. Miguel Chevalier creates fractal works of art using science, computers and aesthetic means.

recreate a composition that puts forward dynamics and movements close to choreographic dance, dense colors whose geometric rigor will oscillate between science and esthetics, fully leading art toward the path of the art sciences: the static comes to life in a vasarelian dynamic. But it is not only abstract art that possesses interpretative exclusivity, since Dürer and Cranach already played with the symbolism of the signs and the space to be interpreted in their respective *Melancholia*, where geometry becomes invisible rather than visible, implicitly organizing the picture and holding its meaning. The same is true of heraldry, which originated and developed in the Middle Ages (Wilhelm 2011b). We have shown elsewhere that artists of the Middle Ages worked on the organization of a geometric coding of colors and space in the representations of the Last Judgments, still spatially identical from the Middle Ages to the beginning of the 20th Century (William 2010b, 2012). This approach, which became an implicit language, made it possible not only to go beyond the usual, more conventional artistic representations, but also to discover true parallel languages, whose alphabets are implicit or explicit colors, numbers and signs that make sense in the same way as any other language.

The works of Roy Lichtenstein are a good illustration of this hybridity process, which integrates the knowledge, techniques and theories of several artists from different cultures to create a unique work. Roy Lichtenstein said of his art "I don't think I make parodies. I think I reinterpret previous works in my own style, like Picasso when he reinvented Velasquez, Delacroix or Rembrandt" (Centre Pompidou 2013, author's translation). In art, it is the hybridity of approaches and theories in contact with other artists, other sensitivities, other spiritualities that has undeniably created innovation and has been a generator of creativity.

Similarly, in theater, more and more productions bring together movement, dynamics, meaning and text, as described by Yves Marc, co-director of the Théâtre du mouvement, who finds himself "at the frontiers of a dramatic dance, an object theater, a textual theater where the body is engaged", linking the *possible dialogues of text and movement*, the body itself becoming geometry in relation to the text:

> How do the "texts" of the body (movement, gesture, composition and body writing) interact with the literary text? How does the latter "incorporate" or not into the fundamental structures of the body? How does this text vibrate in relation to states of thought and emotional states? How can a discontinuous literary text be punctuated and put into perspective by movement? These relationships will be studied in relation to the play of time, dynamics, space, meaning. We will discover the poetic or dramatic gaps that can result from them. (Théâtre du mouvement 2016)

From this kind of approach, it is easy to conceive that the arts can go beyond the stage of simple emotion evoked or perceived, and they can be analyzed in the same way as other sciences and recognized as a *science* in their own right, to be relayed only to the esthetic and emotional sphere of an impalpable and unstable feeling.

In the humanities, geometry is also visually present in the schemes of thought modeling, but perhaps even more so in this case logical modeling, which allows one, schematically, to represent in time and space a theoretical process. Pragmatic semiotics, narrative semiotics, interpretative semantics, discursive semiotics, sign theory and symbolistics have produced numerous theoretical schemes. We think of Peirce and his triads, Gustave Guillaume and his binary tensions, or Greimas and Rastier's semiotic square. When we consider François Rastier, we understand that the semiotic square is a logical square and not strictly speaking a geometric one, but that, on the other hand, anthropic zones can be represented as topological spaces (with "thick" borders), hence a possible projection on mythical narratives. Diagrammatics, the importance of simple forms and symmetries and geometry, especially topological ones, may also have links, but these have yet to be specified.

Without essentializing it or making it an obligatory passage, geometry is a link that is sometimes explicit and direct, sometimes implicit and more distant from our models. Behind the prism of theories, it symbolically links academic contexts: life sciences, physics, mathematics, arts, and human sciences, but also the spiritualities of the world. It becomes, as a common material, a "cement", building bridges between the academic fields and theorizing the cultural sciences, thus offering it a form and a plasticity to be infinitely multiplied. This geometry, at first mainly Euclidean (centered symmetrical figure), becomes non-Euclidean when the theory is deformed by the dynamic process of elasticity (see for the mathematical topology, the Möbius strip).

6.5. Theorists and the evolution of theories

Like cultural objects, theories reflect what cannot be changed without becoming something else. While theories do not transform in the biological sense, at least they are transformed and recomposed. Operating even the smallest change to a theory or a genome undeniably leads to a new theory, a new material, a new creature, a new plastic figure and thus leads to a new thought or creation. This is also true of a vaccine, a text, a painting, any literary and artistic work. The evolution in nature is Darwinian or Lamarckian, but whatever it is, it can be observed in diachrony *a posteriori*. Today, we influence the transformations of certain creatures (plants and animals for the moment[8]) via the genome, but theories, in contrast to matter, genome or DNA, are often associated with a researcher. Theories are registered trademarks, controlled names. They belong, literally and figuratively, to a person, who since antiquity has been called the master, a scholastic term if there is one, because who says master, says pupils or followers. With their theories, the master will found a school, or even a school of thought which will bring a thought to a school[9], in which followers will apply the master's theories, ideology and ethics, until the latter themselves create their own theories in harmony with those of the master or, on the contrary, in open opposition, often a sign of a definitive separation between people.

The thematization of the critical dimension of philosophy goes back to Kant, and further to Platonic dialogics, but philosophy still functions according to the model of adherence or rejection of the above. In the HSS, theories are fairly immutable, for theories and theorists are much rarer than in the life sciences, for example. If there is one field that is evolving slowly and relatively little in terms of theorizing, it is the HSS: to put it simply and briefly, psychologists are still today either Freudian, Lacanian or Jungian, sociologists are pro-Bourdieu or anti-Bourdieu, linguists are pro-Chomsky or anti-Chomsky, creating for decades schools of opposing followers, which does not, however, help advance research, insofar as these quarrels are confined to preserving the theoretical memory of something ancient, almost sacralized. This attitude tends to suggest that too many theories in the HSS are neither elastic nor hybrid: they certainly have a plasticity of their own, but when they block their transformative dynamics, they are then led to a decline because they are not adaptable to the constantly changing phenomena of society.

The scientific approach, which consists of applying only the theories of the past decades, is finally quite medieval because it is pyramidal and vertical. It undeniably represents an obstacle to novelty by preventing more spontaneous, intuitive, even aesthetic (Lichtenstein *et al.* 2013) theoretical productions rooted in feeling or affect, productions that could allow students to create their own theories and explain the various solutions and influences of their own abstract thinking. The opening of theorizing workshops was a step forward in overcoming the medieval notion of the master and their followers, which in some cases could prove to be a stumbling block to theoretical innovation. The best school of thought will always be the one where doctoral students compete in ingenuity under the scientific and benevolent eye of an elder, and not the one where submission to a master and their theories is required, as this hinders any form of renewal of ideas and therefore of thought.

---

8 Even though for at least a decade now, American, European and Australian companies have been trying to patent nature itself, by patenting living things (Les mots ont un sens 2009).

9 In the 13th Century, when Robert de Sorbon created the Sorbonne, he masterfully "taught", literally and figuratively.

6.6. Polysemy of cultural fact and scientific rigor

What also makes the theorization of human and non-human cultural activities complex is the multiplicity of interpretations that can be made of them. Gilles Leguellec speaks of the danger of a semiologization of art due to different approaches and tools:

> But, if I am interested in the exchange of goods, what is to be understood from the point of view of both sides is no longer how the work varies according to the expressions, so that one can say that there are as many objects relative to an image or a work as there are expressions, but the way in which the instructions for use differ and how they are rearranged according to the users. It is no longer a question of semiotics, but of mimetics. [...] To see that there is necessarily a rearrangement of the work, it is enough to compare the material used by the artist with that used by the student; this is obvious when, to analyze a painting, one has only pencils at one's disposal. We will be told that we are not mechanical and that we are capable of proceeding teleotically by manipulating the pencil in chromatization mode and on the other hand of instrumenting, that is to say, of conforming the practice of the pencil to the pictorial model effect. Nevertheless, the difference in device is there, it leads to inflect the response in another direction. In my opinion, this is where the translation (*ductus*) lies, since it is an exchange of conduct that is involved. (Leguennec n.d., author's translation)

Similarly, Gregory Chatonsky speaks of "leafing through" and "leafing through realities" about the symbolic, alphanumeric, iconographic levels of overlapping realities on Facebook:

> [...] it is necessary to see how reality becomes more and more layered, the symbolic, alphanumeric and iconographic levels are superimposed and it becomes impossible to separate them from a remnant that will be the reality in itself. This is certainly not a new question, but here again it takes an explicit social turn. This strata of realities is not without consequences for the artistic imagination: *each event is fragmented into a multiplicity of interpretations*[10]. (Chatonsky n.d., author's translation)

The "strata" notion is interesting. This fragmentation into a multiplicity of interpretations is also what clearly distinguishes the hermeneutic approach of the

10 Emphasis added.

HSS from the so-called "hard" sciences: the arts, contemporary and ancient literature, polysemic or ambiguous advertisements are the best representatives of this. It is well known that each interpreter potentially has their own interpretation, which may moreover be subject to variation over time. Whether figurative or abstract, a painting presents different faces when imaginaries enter into action. The work of art thus escapes its author as soon as it leaves the studio. This is also true of the literary work that becomes that of the reader, the reader ideally hoped for by the writer does not exist, as Umberto Eco has demonstrated as a theorist and novelist (Eco 1999). How to theorize, how to make sense and science in a dynamic context where the imagination constantly changes the interpretative deal? Monodisciplinary theories can take even less account of these multidisciplinary, malleable and transformable phenomena, which are in any case always plural. To take these fluctuations into account is to systematically take into consideration, in theories, the dynamics of the rates of variation and the famous exceptions that confirm so many rules.

6.7. The return of diachrony

The cultural sciences function in diachrony, as do the semiotics of cultures, since the notion of *culture* is intimately linked to that of *time*, development, tradition(s), continuity and separation, overcrowding and distance, permanence and impermanence. Diachrony is one of the main pillars of the cultural sciences (Rastier 2000). Many linguists today claim to be Saussurian in their thinking, yet diachrony is disappearing in universities (Guillaume 2010a). Without diachrony and a return to the cultures and languages of yesterday, it becomes impossible to understand the present. Whether one thinks of research on translations, etymology, lexicology, words, signs, symbols from yesterday to today, on comparisons and cultures, research in the HSS *makes sense* and *becomes science* through the mastery of cultural facts over varying periods of time. All studies on mythologies, beliefs and religions are only diachronic and are only synchronized today by recontextualizing them in present-day societies. The study of the transferability of extra-disciplinary cultural meaning and its theorization is profoundly diachronic, even when dealing with contemporary situations. The work of cultural scientists thus gains in quality by turning again to ancient languages and cultures, not to sacralize them but to recognize their outlines and to recognize them better in their new forms when they reappear in new contexts.

6.8. Conclusion

The artists of the Middle Ages and other cathedral builders gave meaning and form to colors with ingenious architectural processes that are still unequalled. The abstract artists of the 20th Century such as Kandinsky, Herbin or Delaunay made the emotion of the invisible vibrate, offering it alphabets of shapes and colors. Scientists, through technical progress and observation methods, have succeeded in proving and showing the invisible, both microscopic and macroscopic. Architects, by drawing inspiration from art and working on the resistance of materials, have overcome imaginary and technological limits. It is up to the HSS to enter the implicit spheres of intertheoretical creation for more innovation and precision as well as for better recognition of their different fields of reflection and skills.

Recognizing the proximity and comparison of disciplines is already nourishing new, more hybrid thinking, and allows for the integration of new terminology into certain disciplinary fields. The decompartmentalization and hybridization of disciplines as a principle of reflection brings a welcome renewal of thought to theorizations. It is up to the academic evaluation structures of the HSS to adapt to these new ways of thinking, and not the other way round. Today, too many researchers are still penalized because of their inter- or transdisciplinary and intertheoretical approaches, which are an additional asset for the fields of knowledge and for opening up to the cultural sciences and the semiotics of cultures, where science makes sense.

NOTE.– This chapter is the summary of a longer study on intertheoricity which the reader may refer to for further study (Guillaume 2014a).

6.9. References

Ablali, D., Badir, S., Ducard, D. (eds). (2014). *Textes, documents, œuvre : perspectives sémiotiques*. Presses Universitaires de Rennes, Rennes.

Anourhy, G. (2014). *La nuit du vivant. Beauté, Vertus et Délices*. CNRS Images, Paris [Online]. Available at: https://www.youtube.com/watch?v=BkpTbejuCJg [Accessed 18 August 2019].

Art Fractal (2019). La théorie des fractales [Online]. Available at: http://www.art-fractal.com/ [Accessed 18 August 2019].

Centre Pompidou (2013). Présentation de l'œuvre "Bull Profile Series" (1973) de Roy Lichtenstein par Camille Morineau. Exposition Roy Lichtenstein du 3 juillet 2013 au 4 novembre 2013 [Online]. Available at: http://www.dailymotion.com/video/x12bh6j_bull-profile-series-1973-de-roy-lichtenstein-exposition-roy-lichten stein-du-3-juillet-2013-au-4-nov_creation [Accessed 5 April 2019].

Centre Pompidou (2019). Fabrique du vivant Mutations/Créations 3 [Online]. Available at: https://www.centrepompidou.fr/cpv/agenda/event.action?param.id=FR_R-0e1ca28a3e1128d624b2d8375beead0&param.idSource=FR_E-0e1ca28a3e1128d624b2d8375beead0 [Accessed 18 August 2019].

Chatonsky G. (n.d.). Enseignement, "Passage à l'acte : du numérique à l'analogique" [Online]. Available at: http://chatonsky.net/files/pdf/enseignement.pdf [Accessed 1 July 2019].

Chevalier, M. (2017). Complex Meshes 2017 [Online]. Available at: http://www.miguel-chevalier.com/fr/complex-meshes-1?position=12&list=vKL4ig1z5P2iQZeRNFSezO31DhB NyTkJDLIUyfJc4mU [Accessed 18 August 2018].

CIRET (2019). Centre international de recherches et études transdisciplinaires [Online]. Available at: http://ciret-transdisciplinarity.org/ [Accessed 20 June 2019].

Colombat, I. (2017). L'art de la science, la science de l'art. *The Conversation*, 22 November [Online]. Available at: http://theconversation.com/lart-de-la-science-la-science-de-lart-85423 [Accessed 15 July 2019].

Debono, M.-W. (2013). *Écriture et plasticité de pensée*. Anima Viva Multilingue, Paris.

Delahaye, P. (2019). *Des signes pour le dire. Étude sémiotique des émotions complexes animales*. Presses Universitaires de Rennes, Rennes.

Eco, U. (1988). *Sémiotique et philosophie du langage*. PUF, Paris.

Eco, U. (1992a). *La Production des signes*. Livre de Poche, Paris.

Eco, U. (1992b). *Le Signe*. Livre de Poche, Paris.

Eco, U. (1994). *Les Limites de l'interprétation*. Livre de Poche, Paris.

Eco, U. (1999). *Lector in Fabula*. Livre de Poche, Paris.

École Pratique des Hautes Études (2019). Section Sciences religieuses crée en 1886 [Online]. Available at: https://www.ephe.fr/ecole/organisation/composantes/section-des-sciences-religieuses [Accessed 18 August 2019].

Éluard, P. (ed.). (1946). Notre mouvement. In *Le dur désir de durer*. Gallimard, Paris, 83.

ENRA (2019). Performing arts collective [Online]. Available at: http://enra.jp/ [Accessed 18 August 2019].

Greimas, A.J. (1976). *Sémiotique et sciences sociales*. Le Seuil, Paris.

Greimas A.J., Courtès J. (1979). *Dictionnaire raisonné de la théorie du langage I*. Hachette, Paris.

Guillaume, A. (2009). Traduction, sémiotique et praxéologie. In *Penser et Agir : contextes philosophique, praxéologique et langagier*, Alexandre, V. (ed.). Le Manuscrit Recherche-Université, Paris, 395–412.

Guillaume, A. (2010a). Diachronie et synchronie : passerelles (étymo)logiques. La dynamique des savoirs millénaires. *Texto ! Institut Ferdinand de Saussure*, XV(2) [Online]. Available at: http://www.revue-texto.net/docannexe/file/2557/guillaume_astrid.pdf [Accessed 20 August 2019].

Guillaume, A. (2010b). "Medieval" time(s): Last Judgements. In *Semiotics 2009 – The Semiotics of Time*, Haworth, K., Hogue, J., Sbrocchi, L.G. (eds). Legas Publishing, Ottawa, 132–146.

Guillaume, A. (2011a). Vers une symbolistique transspirituelle – pour une laïcité pacifiante. *Cahiers Villard de Honnecourt. La transdisciplinarité, une nouvelle vision du monde*, 80, 103–113.

Guillaume, A. (2011b). Langues et traductions médiévales : Que de mots ! Que de maux! In *Théories et pratiques de la terminologie : analyser des termes et des concepts*, 16, Briu, J.-J. (ed.). Peter Lang, Bern, 131–151.

Guillaume, A. (2012). Dynamiques et formes spatio-temporelles "médiévales" : les jugements derniers. In *Monde(s) en mouvement : mutations et innovations en Europe à la fin du Moyen Âge et au début de la Renaissance*, Cunin, M., Yvernault, M. (eds). Presses Universitaires de Limoges, Limoges, 139–156.

Guillaume, A. (2014a). L'interthéoricité : sémiotique de la transférogenèse. Plasticité, élasticité, hybridité des théories. *PLASTIR, Plasticités, sciences et arts*, 37, 1–34.

Guillaume, A. (2014b). Pour une sémiotique diachronique des cultures : le "Moyen Âge" aujourd'hui. In *Textes, documents, œuvre : perspectives sémiotiques*, Ablali, D., Badir, S., Ducard, D. (eds). Presses universitaires de Rennes, Rennes, 381–406.

Guillaume, A. (2015a). De la morale médiévale à l'éthique contemporaine : sémiotique de la valeur d'hier et d'aujourd'hui. In *Valeurs. Aux Fondements de la sémiotique*, Biglari, A. (ed.). L'Harmattan, Paris, 393–420.

Guillaume, A. (2015b). Intertheoricity: Plasticity, elasticity and hybridity of theories. Part II: Semiotics of transferogenesis. *Human and Social Studies*, 4(2), 59–77.

Guillaume, A. (ed.). (2017a). Traduction et implicites idéologiques, 2nd ed. *Texto ! Textes et cultures* [Online]. Available at: http://www.revue-texto.net/docannexe/file/3909/traduction.pdf [Accessed 12 August 2019].

Guillaume, A. (2017b). Sémiotraductologie et traductogenèse. *Contemporary Issues of Humanities Teaching: Theoretical and Applied Aspects*, 45–55.

Guillaume, A. (2017c). Humanimalisme et lexicologie : les mots de la souffrance animale [Online]. Available at: https://www.youtube.com/watch?v= Fr8vGN4BAEI [Accessed 23 January 2019].

Guillaume, A. (2019a). Le mot sentience entre dans le Larousse 2020. *Revue de La Fondation Droit Animal, Éthique et Sciences – LFDA*, 102, 25.

Guillaume, A. (2019b). La sémiotraductologie : traduire le sens implicite en transfert. In *Au cœur de la traductologie – hommage à Michel Ballard*, Wecksteen, C., D'hulst, L., Mariaule, M. (eds). Presses de l'université d'Artois, Arras, 203–220.

Guillaume, A. (2019c). (Re)Think the human-animal relations: Zoosemiotics and humanimalism. *Human and Social Studies (HSS)*, VIII(1), 13–31.

Leguennec, G. (n.d.). Arts plastiques, sujet et projet. L'anthropologie clinique médiationniste [Online]. Available at: http://gilles.leguennec.free.fr/webquandpuce/artetsociete/theoprat2.htm [Accessed 10 July 2019].

Les mots ont un sens (2009). 20 % des gènes humains brevetés ! La nature en procès [Online]. Available at: http://www.lesmotsontunsens.com/brevet-nature-20-pourcents-genes-humains-brevetes-sante-en-proces-aclu-etats-unis-4565 [Accessed 12 January 2019].

Lichtenstein, J., Maigné, C., Pierre, A. (eds). (2013). *Vers la science de l'art, L'esthétique scientifique en France 1857–1937*. Presses de l'Université Paris Sorbonne, Paris.

Muglioni, J.-M. (1996). Le principe téléologique de la philosophie kantienne de l'histoire. *Kant, Philosophie de l'histoire. Revue Germanique internationale*, 6, 113–127.

Nicolescu, B. (1996). *Une nouvelle vision du monde : la transdisciplinarité*. Éditions du Rocher, Monaco [Online]. Available at: http://ciret-transdisciplinarity.org/transdisciplinarity.php.

Peirce, C.S. (1978). *Écrits sur le signe*. Le Seuil, Paris.

Rastier, F. (1989). *Sens et textualité*. Hachette, Paris.

Rastier, F. (2000). De la sémantique cognitive à la sémantique diachronique : les valeurs et l'évolution des classes lexicales. In *Théories contemporaines du changement sémantique, Mémoires de la société de linguistique de Paris,* IX, Jacques, F. (ed.). Peeters, Louvain, 135–164 [Online]. Available at: http://www.revue-texto.net/Inedits/Rastier/Rastier_Classes-lexicales.html [Accessed 12 August 2019].

Rastier, F. (2001a). *Arts et Sciences du texte*. PUF, Paris.

Rastier, F. (2001b). L'action et le sens – pour une sémiotique des cultures. *Journal des anthropologues*, 85–86, 183–219.

Rastier, F. (2002). Anthropologie linguistique et sémiotique des cultures. In *Une introduction aux sciences de la culture*, Rastier, F., Bouquet, S. (eds). PUF, Paris, 243–267.

Rastier, F. (2003). Le langage comme milieu : des pratiques aux œuvres. *Texto ! Revue de l'Institut Ferdinand de Saussure* [Online]. Available at: http://www.revue-texto.net/Inedits/Rastier/Rastier_Langage.pdf [Accessed 3 February 2019].

Rastier, F. (2005). Doxa et lexique en corpus. Pour une sémantique des "idéologies". In *Du lexique à la doxa. Actes des Journées Scientifiques 2002–2003, Cahiers du Cirlep*, 22, Pauchard, J., Canon-Roger, F. (eds). Presses universitaires de Reims, Reims, 55–104.

Rastier, F. (2006). La Traduction : interprétation et genèse du sens. *Texto ! Revue de l'Institut Ferdinand de Saussure* [Online]. Available at: http://www.revue-texto.net/Lettre/Rastier_Traduction.pdf [Accessed 3 February 2019].

Rastier, F. (2011a). *La Mesure et le Grain. Sémantique de corpus*. Honoré Champion, Paris.

Rastier, F. (2011b). Objets culturels et performances sémiotiques – l'objectivation critique dans les sciences de la culture. In *Performances et objets culturels*, Hébert, L., Guillemette, L. (eds). Presses universitaires de Laval, Quebec, 15–58.

Rastier, F. (2012). Sur la sémiotique: rétrospections ou agenda ? *Texto !*, XVII(3) [Online]. Available at: http://www.revue-texto.net/index.php/http:/www.revue-texto.net/1996-2007/Saussure/Archives/docannexe/file/Archives/Parutions/Parutions/Semiotiques/index.php?id=3053 [Accessed 12 August 2019].

Rastier, F. (2013). De l'essence double du langage, un projet révélateur. *Journal of Romance Studies* [Online]. Available at: http://www.revue-texto.net/docannexe/file/3283/arena_intro_francois_rastier.pdf [Accessed 3 August 2020].

Rastier, F., Bouquet, S. (2002). *Une introduction aux sciences de la culture*. PUF, Paris.

de Saussure, F. (2002). *Écrits de linguistique générale*. Gallimard, Paris.

Société française de zoosémiotique (2019). La SfZ [Online]. Available at: http://societefrancaisedezoosemiotique.fr/la-sfz/ [Accessed 10 July 2019].

Théâtre du mouvement (2016). Les dialogues possibles du texte et du mouvement [Online]. Available at: http://collectifartsmimegeste.com/index.php/fr/actu/collectif/44-les-dialogues-possibles-du-texte-et-du-mouvement-stage-professionnel-du-theatre-du-mouvement and http://www.theatredumouvement.com/index.php?rub=1.

PART 3

Literature and Arts Sciences

7

Challenges of Non-logocentric Semiotics of Cultures: Explorations Based on Music and the Notion of *Significativity*

The logocentrism of the entire Western philosophical tradition, from which arises, among other things, the textualism of the second half of the 20th Century initiated in particular by Genette and Barthes, has often been taken for granted. It is striking, however, that most of the authors who constitute the reference basis of cultural sciences today have defended a non-logocentric approach. Whether Leroi-Gourhan, Simondon, Cassirer or even Saussure – to mention only a few major figures – each time we find the same critical distance from logocentrism, i.e., sticking to a minimal definition, the definition of the fundamental nature of the human being essentially through verbal language.

This critical distancing of logocentrism takes different forms each time: in Leroi-Gourhan's work, what primitively constitutes a human being is bipedalism and the liberation of the hand for the construction of tools, leading to the development of the cerebral capacity to remember and anticipate in an increasingly distant way (thus developing exponentially what F. Rastier calls the "distal zone", the zone of the absent, characteristic of human brain functioning), verbal language being one tool among others to move in this zone. According to Simondon, it is the extreme refinement of a technical thought that characterizes humans; according to Cassirer, it is the ability to historicize mediation types (which he theorizes by the concept of "symbolic forms", in the programmatic plural). Finally, at the beginning of the century, Saussure reminded us that the relationship between linguistics and

Chapter written by Lia KURTS-WÖSTE.

semiology was one of encompassing the former by the latter[1]. Given this observation, what is the specific contribution of the semiotics of cultures that these authors claim to be making today? Certainly the revaluation and redefinition of a certain hermeneutics, which is both non-ontological and is distinguished from a biased idea of structuralism (which assimilates it to a strictly classifying and universalizing thought).

In this context, non-logocentric semiotics refers, on the one hand, to semiotics that considers that the two-plane model of the language sign (valid for the oral as well as for the written word) should not be the archetype of all semiotic functionings, and, on the other hand, a semiotics that considers that the verbal itself cannot be reduced to the "logico-grammatical" (and to its tradition inherited from the philosophical *logos* of the ontological and propositional type) and that it requires taking into account its "rhetorical-hermeneutical" dimension to use the terminology proposed by François Rastier. After recalling the characteristics of such hermeneutics, we propose to explore ways of strengthening its non-logocentric character by relying on the heuristic value of musical hermeneutics, which will ultimately allow us to integrate the material hermeneutics theorized by the semiotics of cultures into the framework of a non-logocentric hermeneutics, which we propose to call "modal hermeneutics".

7.1. Interpretative action, hermeneutic science and the general hermeneutization of the sciences

Seen in the light of the general epistemological framework of the cultural sciences, the human and social sciences are today becoming more aware of the particular quality of their mode of scientific objectification. Furthermore, they are gradually acquiring the means to claim a dignity at least equal to that of the logico-formal sciences for the cultural objects that interest them – primarily works of art – insofar as the latter require specific restitution procedures. Their status as hermeneutic sciences is thus reasserted, if we take up the typology proposed by Jean Ladrière (2005), which distinguishes three kinds of sciences: formal, empirico-formal and hermeneutic.

The semiotics of cultures, of the Cassirerian tradition, thus aims to reconstitute a "family" of critical hermeneutics, adopting a comparative and historical approach (in

1 This simple reminder invalidates Derrida's criticism of Saussure's logocentrism based on a biased and non-specific interpretation of his writings, which links him to the ontological linguistic tradition via phonocentrism and the supposed ideal of "presence" (whereas Saussure based his entire semiology on difference, i.e. the definition of a unity by the *absence* of all others).

the sense of taking into account the corpus and the forms and norms of the tradition), which have appeared, starting in the 18th Century in order to prolong their efforts:

> In order to sketch its outlines, it would be necessary to *re-examine* [our use of italics] the works of Levi-Strauss and Marshall Sahlins in anthropology, those of Cassirer in philosophy, those of Dumézil and Charachidzé in comparative mythology, those of Carlo Ginzburg in history, those of Panofsky, Saxl, Gombrich in iconology, those of Segre in philology, those of Lotman and Uspensky in semiotics. (Rastier 2011, p. 15, author's translation)

This without neglecting their differences, as shown, for example, by Muriel van Vliet's question concerning the potential distinctions to be drawn between Panofsky's approach and that of his illustrious predecessor:

> How did Panofsky develop the iconological method of interpreting works of art based on the notion of symbolic form and indirectly, via Cassirer, on Humboldt? Does he distance himself from Cassirer in this respect? Some of these questions have been addressed in the research literature, notably by Sylvia Ferretti$_3$ and Michael Ann Holly; Audrey Rieber's recent work contributes to putting these approaches into perspective and discussing them. (van Vliet 2016, author's translation)

But beyond the necessary joint work of distinction, through the recognition and strengthening of the definition of an interpretative *action* specific to the hermeneutic sciences, it is generally a question of: (1) compensating for a form of collective amnesia that has caused us to lose contact with such projects of defining hermeneutic scientificity, whether this amnesia is the consequence of the great current of scientific positivity centered on "data" and the logico-formal approach or, conversely, the consequence of currents that consider subjective experience as the priority, or even exclusive, criterion of truth; (2) to claim at the same time the legitimacy of a plural conception of scientific rationality and the interest of a bringing together of the hermeneutic sciences in the same federative project – the tradition of such a hermeneutic science being well attested, as the list of authors quoted above shows, even if it has sometimes been denied by the ideal of a totalizing Big Science. The model of a scientific hermeneutics, which is moreover applied to works of art, is however far from being a consensus and must be defended on at least two fronts at the same time: its very legitimacy is questioned at the same time: (1) by the aversion to science, characteristic of the Heideggerian existential hermeneutics, according to which science "does not think"[2] (only philosophy

2 Gérard Dessons notes that while Heideggerian hermeneutics "thinks", it is in the mode of the exegesis of the sacred texts, whose vocation is to return to the anagogical meaning beyond

"thinks", because only philosophy thinks about "being" in its original relationship with *Grund*); (2) symmetrically, by the scientist's aversion to hermeneutics, in the spirit of which the tradition of a scientific hermeneutics and the tradition of an ontological and existential hermeneutics are undoubtedly not differentiated enough. Two antagonistic positions to which one would probably have to add the general prejudice that a scientific approach to the arts would cause them to lose what is most precious, namely their intensity – moving them irremediably away from the question of taste and value, which are central to an anthropology of art.

François Rastier, in the field of linguistics applied to literary texts, has decisively contributed to defining interpretative action as a *praxeology* that strives to characterize works in their diversity, within the practices that constitute them and against which they are measured, while recognizing the interpreter as an actor endowed with affects and responsibility.

The dimension of *action*, from which every work results, makes it possible to conceive about cultural and, more specifically, artistic practices according to a transformational energy that invites us to restore "in the theory of art, questions of craftsmanship and quality of execution, such as the material singularity of the object or artistic performance – in its smallest details, as in a recital, the tuning of the instrument, the acoustics of the room, the positioning of the performer, etc." (Rastier 2016, p. 218, author's translation). Such a position is inspired by Simondon's approach, according to which the work is supported by "the underlying solidity of technical judgment to sustain it: the work of art is a thing that has been made" (Simondon 2017, p. 205).

Such an approach overcomes the school-like quarrels in text sciences, for example between discourse analysis (which favors the non-specification of the work of art) and stylistics (which has often essentialized "the literary") to place itself on an anthropological level, by placing the concept of work at the heart of the project of an *operatic* or science of works. While the theories of enunciation account for language as an act by reference to the intentionality of the speaking subject, whether

the "literal meaning": it is from the "relationship between the letter of the sacred scriptures and the speculative thought of theology" that Heidegger elaborated his conception of poetry. Similarly, Jauss constitutes literary hermeneutics concerning Gadamer's proposal to "redetermine the hermeneutics of the humanities on the basis of juridical and theological hermeneutics". His method of discovering the *significativity* of a poem through the exercise of a two-stage critique – a first perceptive moment preparing the second interpretive moment – is done exactly on the model of the exegesis of sacred texts, which makes a literal and an anagogical moment follow one another (Dessons 2004, author's translation).

conscious or not, an operatic aims to stick to the semic level, conceived as the result of an interpretative journey that takes into account multiple parameters (type of text, genre, style, general esthetic project, corpus, etc.), so that Rastier can praise the salutary absence of the theory of enunication in Saussurean semiology.

On the receiving side, interpretative praxeology is specific in that it simultaneously combines the three types of human activity – human *action* – theorized by Aristotle: (1) construction/manufacturing (*poïèsis*), insofar as the interpreter must keep in mind the constructive dimension of his or her reception, in the sense of an acute awareness of the fact that "data are what one gives oneself"; (2) to know it (*theôria*), in this mixture of knowledge and understanding characteristic of Cassirer's "expanded kanticism" (where knowledge is conceived as a forgotten interpretative action); (3) to act (*praxis*) in its ethical dimension – interpretative activity, controlled by a deontology, having a clear dimension of responsible *action*: such praxeology goes beyond a simple "praxematic" approach to language insofar as it consists of an awareness of the social, and even ethical, political role that the criteria that regulate interpretative practices may have – we shall return to this last point at the end of this chapter.

On the other hand, in the field of semiotic studies, the generative perspective adopted by Greimas and, after him, the Paris school to define interpretative *action* paradoxically avoided having to formulate a theory of interpretation. In fact:

> The deep structure concept presupposes a very strong hypothesis: that the explanatory model is immanent to the object described. Deep structures easily accommodate various philosophical actors, from the human mind according to Chomsky to the Lacanian subject according to Kristeva. (Rastier 1987, p. 219, author's translation)
>
> In short, it seems that contemporary semiotics and linguistics have not been able or willing to produce a theory of interpretation for three interrelated reasons: (1) their deductive character (Greimas, Hjelmslev); (2) their generative character (Chomsky, Greimas): the text is the arrival point of generation (and not the starting point of interpretation); (3) their formal character, which led Hjelmslev and Chomsky – whatever their differences – to reject semantics, and then belatedly to grant it an ancillary function. (Rastier 1987, p. 217, author's translation)

While interpretation is no longer based on a vericonditional logic, on a deductive or generative logic, nor on a formal logic, it is still possible to verify its consistency both by the clarity of its criteria and by the sum of its distinctions: the science of values and not of facts, of conditions and not of laws or foundations *a priori*, of the certain and not of the true, of norms and not of rules, of relations and not of substances, of the singular and not of the universal. One could say that it thus becomes a *science of objects that resist*, in the sense that one learns from works how to interpret them and recognizes the constraint of linguistic material on interpretation.

With the firm definition of such praxeology, the so-called "soft" sciences are a little less so, finding today more than ever the means to decomplex themselves in the face of a supposedly unique model of scientificity, by proving the solidly-criticized, even if still fallible, procedures they can rely on.

Certainly, as Marie-Dominique Popelard and Denis Vernant remind us:

> No human science yet has the rigor, the level of formalization and axiomatization of the natural sciences. Hence the current division between the "hard", noble and indisputable sciences and the so-called "soft" sciences, which some even propose to call "tender" to avoid any pejorative aspect! (Popelard and Vernant 1997, p. 84, author's translation)

However, the semiotics of cultures responds, on the one hand, that, in terms of rigor, it has the means to verify its capacity to be a science of the certain (and not of the true) and, on the other hand, that, as far as the "level of formalization and axiomatization" is concerned, the latter is not necessarily a desirable or unique objective if we want to guarantee its capacity to characterize the singular – which removes axiomatization from its status as a criterion for global evaluation. On the other hand, the recognition of the interpretative action as a responsible act leads conversely to conceive the interest of a general hermeneutization of the sciences: in this sense, all the sciences are human sciences in the sense of sciences *made* by man. They present common characteristics, peculiar to human symbolization: mediations, the importance of the distal zone (of the zone of the absent; for example, of previous theories), the importance of the dimension of "symbolic debt" toward the great historical referents, stylization within a tradition (stylization that characterizes human activity as a critical effort to eliminate the superfluous, rather than as an addition of data, a characteristic that testifies to an exceptional level of complexity, of which machines would not be capable), transmission, diversification, etc.

The reconception of semiotics as hermeneutics and hermeneutics as praxeology could thus serve as a compass for the cultural sciences in the making today, as well as anthropological and cultural-based "spectacles" through which to enhance some of the major theoretical and practical achievements of the human and social sciences in terms of interpretative capacity – so that the hard sciences could gain through them a better awareness of the symbolic stratifications used to account for the phenomena they describe. This is what Nelson Goodman reminded us of in *Ways of Worldmaking,* and it is also what Bruno Latour reminds us of today, in his *An Inquiry into Modes of Existence* (2013) or in *Jubiler ou les tourments de la parole religieuse* (2013) – whatever one thinks of his views, which are cavalier to say the least, even reactionary[3], on artistic practices and his proposal to give origin to human practices and human *know-how*, elsewhere than in the human being[4]:

> One must never have touched the length, the precariousness, the splendor, the originality of the cascades of references that allow an astronomer to access the confines of the big bang, an oceanographer to map the movement of tectonic plates, a mathematician to follow the proof of a theorem on the theory of numbers, to a historian to recompose the traces of a popular uprising ignored by all, to a [hermeneutist to characterize the singularity of a work to] believe that the world left in the wake of science would be basely material and visible, objective and stubborn, simply and stupidly there. If one really holds to the adjective, what could be more "spiritual" on the contrary than the worlds born in the process of the sciences? What, in any case, of *directly* visible? (Latour 2013, p. 40, author's translation)
>
> The sciences, in the plural, the real ones, the ones whose laboratory life can be studied, the teams, the equipment, practice a risky operation: each production undergoes a painful transformation since the statement never resembles what it refers to. But in double-click communication, all difficulties vanish, all paths flatten: information

3 "[…] art is too mysterious, too spiritual, too much worked by the beyond, too enigmatic, too innovative, too perverse also to accompany the path of religion for a long time. It disturbs, it invents, it attracts, it does not arouse the people of those who understand the tradition *that we have* been interpreting until then without understanding" (Latour 2013, p. 122, author's translation).

4 "We only keep, in order to get along, what comes out of our hands without them being able to believe the origin" (Latour 2013, p. 165, author's translation). But what Bruno Latour then says about the religious word seems to us to be able, without cynicism and if we shift his remarks to a semio-anthropological level (making him lose, we are aware, his religious stakes), to have an interest in *science*, in the valorization of its constructive character, which it is a question of understanding that it does not make him lose the joint aim of rational objectification: "Is it real or is it fabricated? I have to be able to say 'both'. And let my opponents, like those who think they understand me, not conclude that I take refuge by this answer in the cynical illusion of a maker of idols who attributes to the productions of his hands the reality he has projected on them" (Latour 2013, pp. 165–166, author's translation).

> becomes faithful communication without any transformation, simply because of the obvious resemblance between the copy and the original. Pure phantasmagoria of course: no science would be possible by imitation, transparency and fidelity. And yet, because of a history that has no place here, this type of communication under the name of "Science" has taken the place of *the* sciences, suddenly concealing their prodigious transformations. (Latour 2013, p. 30, author's translation)

In this appreciation of *action* – of the awareness of not only the agentive but also the self-reflective character of this action – one can recognize the legacy of the Saussurean tradition as it is today re-established by the important philological work carried out in recent years. The Saussurean reflection on language and linguistic practices concerns the question of *action* at least in two ways: from the point of view of the genesis of signs in speech and from the point of view of the linguist's critical reflection on what he does – and therefore on the objects he gives himself. The Saussurian systematics based on the principle of a *dual essence* of linguistic facts is a systematics that has taken note of the fact that, as Simon Bouquet reminds us, "the field of knowledge (and that of the theory of knowledge) is [now] determined by the conceptual field of action (and that of the theory of action)" (Bouquet 1997, p. 33, author's translation). Thus, the thought of the Saussurian sign as the result of an action (an artistic action and a theoretical action that attempts to describe it) is at least as important an aspect, in terms of language semiotics, as the pragmatic recognition of the act dimension of saying. Starting from Saussure, metalanguage cannot do without a reflection on the responsibility of the linguist in the construction of the language object – in order to better define this agentive, dynamic, transformational dimension of language:

> In short, the genesis of signs through the particular semiosis of texts continues and concretizes that of language. As it is a praxeology, this dimension has not been reflected by the grammatical tradition, and the persistence of philosophical ontologies in contemporary linguistics has hardly allowed us to go any further. Pragmatics has certainly been concerned with what we do with language, but not with how we do it. (Rastier 2018a, p. 3, author's translation)

The semiotics of cultures insists on this aspect, from the perspective of a Humboldtian energetics that is itself the successor to Goethe's metamorphic vision, which leads to thinking of the linguistic sign less as a unit than as a "passage", a definition that decisively separates the sign from its canonical definition as a predefined unit within a nomenclature indexed on *realia*.

The need for a praxeological approach to interpretative action is particularly evident, as we have already noted, in the study of the arts: it is a question of

(1) recognizing the specificity of this corpus; (2) admitting the interest and legitimacy – as well as the difficulty – of a science of the arts which can take advantage of interpretative principles applicable in a transhistorical and transcultural manner, which leads to the simultaneous admission of the interest and legitimacy of a science of the singular, concerned with a time of tradition from which works are elaborated, in contrast to the Aristotelian definition according to which there is only a science of the general: it is indeed the hermeneutic *principles* that are transhistorical, while it is a question of restoring the irreplaceable specificity (historical, cultural) of the objects that it sets out to study.

If the epistemological framework of a hermeneutic science is thus salutarily strengthened, the question of the possibility of describing all the components and semiotic levels of a work is problematic and cannot – this is our position – be limited to a semantic theory.

7.2. Hermeneutics of non-verbal objects: a challenge for the semiotics of cultures, a benefit for thinking about the reinsertion of a theory of meaning into a theory of stakes

A science of the arts, in fact, is not only concerned with the language arts; it is even traditional to separate "literature" and "arts", as if the latter should only concern the non-verbal arts. The cultural sciences are therefore of course intended to restore the singularity of non-verbal cultural objects, in keeping with the position adopted by Cassirer:

> If it is true that art cannot be taken further without a linguistic approach and that language cannot be understood in all its depth without an artistic approach to culture, it is never a question of reducing art to language or language to art in Cassirer's work. (van Vliet 2016, p. 1, author's translation)

For the so-called "visual" arts (but is not written language art just as much, if not more so?), the list of hermeneutics is long and well attested, even if the question of the mode of construction of meaning upon reception of images remains a controversial issue in the semiotic field[5].

For our part, we propose to take music as an example according to the semio-anthropological perspective specific to the semiotics of cultures – without claiming to be exhaustive, but rather as a heuristic experiment – hoping that this detour will yield a double advantage: (1) for the conception of the relationship between musical

5 For an elaboration on this issue, see, for example, Dondero (2019).

studies and cultural sciences; (2) for the elaboration of the definition of the regime of semioticity specific to non-verbal cultural objects.

1) With regard to the relationship between music studies and cultural sciences, we would like to contribute both:

1a) to a better recognition of the importance of music studies in the field of human sciences, admitting with Jean Molino that:

> The root of the error is, at the bottom, to believe that language is the model for all symbolic phenomena. In this, the study of music makes an essential correction and contribution to the knowledge of the symbolic [...]; (Molino 2009, p. 88, author's translation)

1b) symmetrically, to a strengthening of the awareness by musical studies themselves of belonging to the cultural sciences and a better ability to orientate oneself and to evaluate the relevance of the different currents of musical analysis, in particular those claiming to be hermeneutical, which remain very heterogeneous, as Gianmario Borio points out:

> To stick to American musicology, the term "hermeneutics" can be applied to esthetics in general, to the method of deconstruction, to post-modern thought, and is often mixed with a critique of society of Marxist origin. (Borio 1996, p. 87, no. 18, author's translation)

2) From the point of view of a semiotic reflection on the regime of elaborating meaning for non-verbal cultural objects, it is a question of considering:

2a) not that music is always asemantic[6]; it is enough for us that it is sometimes asemantic – for example, in the case of a piece for an instrumental ensemble without

---

6 As we know, no consensus has been reached on this issue, perhaps because of a lack of in-depth reflection on the common vocabulary to be used in order to reach agreement. On the one hand, Stravinsky's or Boulez's position can be adopted, following that of Hanslick: Since the famous *Vom Musikalisch Schönen* (*From the Beautiful into Music*) [1854] by Eduard Hanslick, most of the innovative composers of the 20th-Century adopted Stravinsky's position: "Boulez asserted without concession: 'Music is a non-significant art' (Boulez 1985, p. 18, author's translation)" (remarks quoted by Jean-Jacques Nattiez, from the article "Analyse et sémiologie musicales", in the *Encyclopaedia Universalis*). On the other, there are Martha Grabosz's assertions about the need to think in terms of musical semantics. Both positions seem to us to be erroneous, both because of their radicalism (there are many cases of musical symbolism, ritual in particular) and because of the lack of a clear theoretical position with regard to a linguistic semantic theory. For a discussion of musical semantics, we refer to the canonical works of Jean-Jacques Nattiez, in particular (Nattiez 1975). In any case, as Robert S. Hatten rightly notes, "Agawu's topical designations, or Tarasti's and Grabosz's semes and classemes can be reproached for being on the one hand too loosely applied – in

any particular ritual or iconic designation – to consider the need to find an adequate vocabulary, capable of highlighting the way in which it can continue to concern "life in meaning", as Cassirer put it, as a cultural object;

2b) the retroactive interest of such a non-logocentric definition applied to music in order to approach the semiotic functioning of the verbal itself and other non-verbal semioses.

It is thus laying the foundations of a musical hermeneutics that can claim, without necessarily saying so explicitly, an interpretative praxeology integrated into a culturalist semio-anthropology. As far as the musicological field is concerned, Jean Molino identified the first lineaments of a musical hermeneutics of this type toward the end of the 19th Century in Germany:

> Musical analysis in the strict sense of the word was born with the *Führer durch den Concertsaal* of Kretzschmar (1886) and the works of Riemann [...]. From its birth, it manifested an ambiguity that stemmed from its origins [music criticism, resulting from a change of perspective, which focused on the study of reception during the 18th Century, as evidenced by the emergence of the aesthetic][....] With Riemann […] a more technical analysis is constituted, the *satztechnische* or *formaltechnische Analyse,* whose project is to give an account of the entire score, from the bottom – the smallest note – to the top – the overall construction of the work. Riemann's own work was partly distorted by the *a priori* he imposed on analysis as a consequence of his harmonic and metric theories. But it is clear that it is in the tradition of Riemann, and in particular in the formal analyses of the German school, that one comes closest to the requirements of a rigorous and explicit method of analysis of musical works. (Molino 2009, p. 98, author's translation)

It is not by accident that the characteristics of Schleiermacher's material hermeneutics, constituting a major reference for the semiotics of cultures, were defined at the same time and in the same geographical area. This is confirmed by Robert S. Hatten, who noted that in Europe "the literary hermeneutics of the 19th Century inspired a tradition of musical hermeneutics" (Hatten 2002, p. 568, author's translation). This theoretical synergy corresponded to the emergence of the general frameworks of what would soon become the semiotic paradigm of the first structuralism, within an extremely demanding philosophy of culture insofar as it refused to endorse a radical break between scientific culture and literate/artistic

contrast to the problem encountered in the paradigmatic analysis by Ruwet and Nattiez – and on the other hand hardly interpreted" (Hatten 2002, p. 570, author's translation) – which makes Grabosz's proposals fragile, to say the least.

culture, i.e. between knowing and understanding. Such a hermeneutic approach allows us to define interpretation as a *strategy* for adapting analytical tools to the specificity of the object. This approach makes it possible to think precisely in terms of a logic that is both federative (because it is centered on symbolic forms) and differential (because it is concerned with characterizing works in their specificity, without considering the verbal as an inescapable model). It is a fruitful alternative to attempts at formalization in musical semiotics, as Jean-Jacques Nattiez, for example, was able to produce with his paradigmatic analyses, whose limits were quickly seen due to their inability to define the singularity of a work:

> [...] paradigmatic analyses often produced an overabundance of detail that subsequently made poetic and esthetic analyses somewhat difficult. From a practical point of view, one wonders why potential meaning was not considered from the outset. (Hatten 2002, p. 567, author's translation)

While the relevance of the notion of "potential meaning" remains debatable, it is nonetheless true that Jean Molino seems to be right to see in this type of "neutral" analysis a form of scientific illusion, under the influence in particular of linguistics, which may have confused science and formalization for a time, whereas the formalistic approach cannot take into account two major data of hermeneutic-type scientific description: diachrony and corpora.

Although he uses it essentially to describe verbal objects, François Rastier has proposed a collection of descriptive terms that are epistemologically important because of their non-logocentric potential: "content", "scope", "address", "destination"; and for the definition of metalanguage itself: "point of view" and "guarantee".

To give just a few defining elements[7]:

1) "Content" refers to (i) the material (which is not reduced only to notes, signifiers or colors, but also concerns forms inherited from tradition); (ii) the degree of elaboration; (iii) the quality of execution. This definition of "content" can be related to that of musical material by Jean Molino, which also includes the question of execution:

> Music involves work on materials, vocal sounds and instruments, thanks to techniques corresponding to knowledge that is more or less explicit; it is this anthropological grounding that constitutes the foundation of "pure" music in the second sense that we have distinguished that of the [only sound material used in musical

7 For more details, see Rastier (2018c, pp. 107–108).

> activities, which corresponds to any kind of *emic* musical entity whose combinations allow musical production]. (Molino 2009, p. 277, author's translation)

The *emic* and not just the *etic* aspect of such material (i.e. the question of the relevance or value of the material elements with regard to the cultural tradition of which they are part) should not be underestimated, as Jean-Jacques Nattiez reminds us:

> A native African plays a tune on his bamboo flute. []...then the native plays the same tune on another flute. The European thinks it is a different melody, because the pitches of the sounds have completely changed due to the construction of the new instrument, but the native swears that it is the same tune. The difference is that the *most important thing* [our use of italics] is the timbre for the native, whereas for the European it is the pitch of the sound. The important thing in music is not the natural given, it is not the sounds as they are made, but as they are intended. The native and the European hear the same sound, but it has a completely different value for each, because their conception is based on two completely different musical systems; sound, in music, works as part of a system. The achievements can be multiple, the acoustician can determine this exactly, but the essential thing in music is that the piece can be recognized as identical, that [is to say that the dimension of relevance is taken into account in the definition of the "material", and not only the strictly acoustic aspect]. (Nattiez 1975, pp. 197–198, author's translation)

2) The "scope" of a work is assessed within different contexts: its artistic scope is assessed on the basis of the corpus from which it is elaborated and the one in which it intends to appear. Its social scope depends on the link between its ethical legitimacy and the esthetic engagement manifested by its project. For example, when Mozart got the German language accepted for opera, it was of such social scope and revealed the paradoxical insertion of the works into a tradition constitution with disruptions.

The notions of "address" and "destination" complete this descriptive vocabulary by taking into account the inscription of the reception within the works, which also intervenes, as we have just seen, in the evaluation of the scope of a work. Thus, the distinction between destination and address, regrettably ignored by the theories of communication, is nevertheless a founding duality of the communicative intention, and is part of the problem of exchange (or even gifting), and, by extension, of transmission.

To complete this vocabulary, it seems important to us to integrate another notion that is of interest in order to identify both the specificity of the semioticity regime for non-verbal objects and the specificity of the interpretative strategies specific to a science of the singular: the notion of the "stake". This is a non-specialized notion, which belongs to common vocabulary, but whose importance in semiotics seems to us to have been underestimated up to now. Such a notion seems to us to deserve an elaboration to serve nothing less than a reintegration of material hermeneutics into modal hermeneutics: taking music as an example allows us to highlight its interest. The *Dictionnaire historique de la langue française* defines the term "stake" as a "risky sum in a game" (c. 1370), which has become commonplace in the figurative sense of "what one risks to win or lose in an enterprise" (1798).

Thus, it is only when content is linked to scope that the issues of a cultural object can be assessed; but, as we have seen, the description of content is itself dependent on the assessment of the scope and stakes of the physical elements. In the case of a painting, for example, if one considers only its content, i.e. its material presentation, then "one sees nothing" (Arasse), in the sense that it is the evaluation of the degree of stakes contained in such a pictorial configuration that makes it a salient material element to be retained for description. Thus, "content", "scope" and "stakes" are in a relationship of mutual dependence and imply continuous adjustments.

In fact, with the question of *action* – of artistic action *and* interpretative action – the notion of the "stake" seems central to us: it is a question of both questioning what a work risks (to gain or to lose) *by doing* and, jointly, of what the hermeneutist risks (to gain or to lose) by their interpretative action.

From the perspective of a cultural semio-anthropology, what does a work want to gain if not the elaboration of an irreplaceable world? And what does it risk, if not being inconsistent?

Symmetrically, what can the hermeneutist want to gain, if not the capacity to restore the singularity of a work, in its relationship with the bodies with which it competes? And what does he risk losing, if not specificity and intensity?

As we can see, the very formulation of these questions is only possible within the epistemological framework of a semiotics of singularities, which is in no way contradictory to taking into account the time of tradition(s) and the collective scope of artistic practices and the objects that result from them: by enhancing the notion of the "stake", we would like to make sensitive the not only scientific but also ethical force of such a position.

7.3. *Significativity*

The notion of the "stake" makes it possible to approach the particular semiotic regime for elaborating the meaning of non-verbal objects (in the sense that they do not meet the Saussurian definition of a "double essence" and do not present a linguistic signifier/signified duality) by making it necessary to mobilize a notion distinct from "sense" and/or "signification": that of "significance".

Such a notion has nothing to do with the technicality of related terms, such as "*significance*" for example, which has not left the linguistic/semiotic field[8]. It is indexed neither in the *Dictionnaire des sciences du langage* (2004) by Neveu, nor in the *Dictionnaire raisonné de la théorie du langage* (1993) by Greimas and Courtès, to take two reference works. The definition of *significativity* can be shared by all, and it is also used in recent work in pharmaceuticals, with the meaning we wish to give it: *Étude de la significativité clinique des interactions entre médicaments du système nerveux central (sédatifs) et spécialités contenant de l'alcool* (thesis by T. Tripon, defended at the Université Paris Descartes in 2007). *Significativity* is the substantial derivative of the adjective "*significative*", which is defined in the *Dictionnaire historique de la langue française as* follows:

> The adjective has detached itself from the general value of meaning, qualifying (c. 1560) that which clearly expresses something, hence (1694) that which reveals or is very characteristic. By extension (1870), it also corresponds to "noteworthy, important", then opposed to *insignificant.*

It is this last meaning that we will retain: it is a question of approaching the meaning of a verbal, pictorial, musical phenomenon, etc., not first of all by defining its meanings, but by evaluating its *importance* (the two being linked, of course, when there are signified elements).

In more technical terms, this definition of *significativity* is superimposed on the semantic dimension for verbal systems, but we would like to emphasize that in law

---

8 We will not go into the complex history of historical variations in the definition of the related terms "meaning" and "sense" in the field of linguistics. We will stick to the definitions proposed by François Rastier: "meaning" concerns lexicographical studies, according to an atomistic and decontextualized approach, while "sense" concerns contextual signified elements actualized in a text, the results of an interpretative process. "Meaning" and "sense" are therefore conceived as passages of texts, some of which are grounded by usage and archived in dictionaries, while others are renewed with each reading, according to historicist, comparative and philological criteria.

it can receive specific treatment and attention to enable thinking about the arts together, both verbal and non-verbal.

Such a regime makes it possible to define "content" (including semantic content) as the result of an interpretative process that justifies its choices by evaluating the importance of the aspects selected in the process of individuation/singularization that is proper to any artistic endeavor.

It should be noted that the considerable development of musical analysis in the 20th Century was accompanied by a claim to the analyst's right, even duty, to question the "meaning" of a work, as is shown, among many others, by this remark taken from the *Guide pratique d'analyse musicale* (1991) by Naji Hakim and Marie-Bernadette Dufourcet[9]:

> The musician must learn to free himself from the indispensable stage of simple observations, to go and touch the *profound meaning* of the piece being analyzed. Any theoretical work [...] only provides tools whose resources must be discovered in practice. (Hakim and Dufourcet 2001, p. 15, author's translation)

If the formula is highly problematic, we must not be too quick to dismiss the musicologists' "concern" with meaning, and recognize with them that if music were thought of as *insignificant* matter, it would not have the force that animates every work of art: the question is therefore not null and void, but it deserves to be rephrased: the notion of "*significativity*" – of which the *Dictionnaire historique* makes the defining antonym of "insignificative" – can contribute to this, it seems to us.

In what way does a musical work participate in life in terms of meaning? According to this modal-type semiotic approach, it participates in it in a principled way (beyond the possible consideration of a combined ritual or iconic symbolic meaning) as it claims to make itself *significant* in the cultural field.

Seeking the significant aspects of a singular work (aspects that are not necessarily *elements*, according to an atomist perspective, but sometimes also general aspects, a tone, a particular management of rhythm, etc.) is therefore to put the very possibility of describing the content under the condition of the explanation of its stakes, multiple stakes, at the same time esthetic, ethical, political, metaphysical, etc.

---

9 See, for example, Meyer (1956). As also evidenced by the *14th Congress on Musical Meaning* organized by Eero Tarasti and Oana Andreica at the "Gheorghe Dima", Academy of Music, Cluj-Napoca, Romania, May 11–15, 2018.

With the notion of "*significativity*", the interpretative act of describing and restoring what is irreplaceable about a work becomes indissolubly linked to the following question: how can we legitimize the choice of highlighting this or that element?

Such a notional couple ("stake"/"*significativity*"), *a priori* rather commonplace, thus has the advantage of fundamentally integrating into hermeneutic practice another notional couple: the duality of "point of view"/"guarantee" and of raising musical studies to the level of a hermeneutics of value – thus echoing the recent and salutary reintegration of the question of value into literary studies themselves (Vaugeois 2006; Rabaté 2007).

7.4. Music and the hermeneutics of *significativity*

It seems interesting to us to reread certain aspects of Adorno's musical analyses in this light: for example, as Gianmario Borio reminds us, "the categories of analysis [proposed by the latter] are drawn from the horizon of the problems to which the work was intended to respond" (Borio 1996, p. 89, author's translation).

The analysis thus focuses on "understanding the critical nodes" (which *we* interpret as the designation of the most significant aspects). The analyses of Schönberg's *Phantasy for Violin with Piano*, but also of Webern's *Six Bagatelles*, for example, are based on the definition of a powerful stake, that of the crisis of musical form: "In each section *something serious* happens [our use of italics]. The technique is that of permanent damage" (Adorno 1963, quoted by Borio 1996, p. 90, author's translation). Thus, *significativity* considers life in the sense in terms of weight – gravity, precisely. Gianmario Borio describes Adorno's analyses themselves as "very significant", according to a kind of recursivity that is quite understandable within modal hermeneutics: as it engages its author, it lends itself in return to a qualitative evaluation. Here, we find the implicit argumentative aspect of a hermeneutics of *significativity* defined as a responsible praxeology. Thus, according to Adorno, "Berg's work of art wants to unify two absolutely contradictory things: to dissolve and to continue to be fully in control of oneself". This is a particularly powerful and difficult stake, and one that is intensified by the perception of this said stake:

> Adorno's method of analysis for interpretation ["for the interpretation": here, in the sense that analysis prepares the musician to interpret, to play a work on his instrument, on the other hand,] allows access to an intermediate region where the analytical criteria are based on the structure of the work on the one hand, and on the

> other hand can be transformed into live enjoyment. (Borio 1996, p. 89, author's translation)

This "intermediate region" seems to us to be that of *significativity*: it presents the major interest for us today to combine "structure" and "live enjoyment" – in short, to reconcile what was disjoined in the 1980s on the grounds of a claim by "structure" to be only an "autonomous entity of internal relations" (Hjelmslev).

A hermeneutics of *significativity* engages a position that is jointly historian, comparative, materialist and responsible, a kind of postural hapax that the academic boundaries between language sciences and literature make non-existent, whereas it has been and continues to be used in art history, its semiotic takeover by the notional couple "stake"/"*significativity*" having the merit of underlining its non-logocentric potential. Modal hermeneutics can thus be a fruitful alternative to the various options already mentioned, i.e. a non-specifying hermeneutics, a desiring and projective hermeneutics and a universalizing hermeneutics, which would think, through the use of "profound structures", that the explanatory model is immanent to the object described.

In this way, Adorno succeeds in formulating new formal concepts, based on his knowledge of earlier compositional techniques, to designate what he calls the "material morphology" of the works of Schönberg or Webern: "dissolution", "intonation", "interrupted time", etc. (Adorno 1963).

One could also take up, but this would require specific work that is not conceivable in the present work, the distinction made by Carl Dalhaus (1990), one of the musicologists most sensitive to hermeneutic questions and their relation to the concept of "value", between *Werturteil* (value judgment) and *Sachurteil* (factual judgment), through which the question of an approach to music through *significativity* would undoubtedly be possible.

While the specificity and the interest of such modal hermeneutics seems to us to be able to be highlighted in a privileged way through the study of instrumental music, it has no reason to remain specialized in this field, and on the contrary reveals its interest within the framework of a self-reflection of material hermeneutics, including applied to verbal works. If we thus attempt to reinterpret an interpretative semantics project by this measure, "*significativity*" opens up the following questions: what is at stake in this or that semantic organization? Let us take the example of the "*fatrasies*" in *Waiting for Godot*, a work characterized by its laconism: while the question of their interpretability can no longer necessarily be evaluated in logico-grammatical terms in their relationship to grammar and logical acceptability, such realizations persist in concerning life in the sense of raising the following questions: in what way are they significant with regard to the overall esthetic project? In what way do they retain a cultural interest, particularly in their relationship with traditions that they both mobilize and transfigure (medieval "*fatrasies*"? Shakespeare? Lewis Carroll? Symbolist or surrealist theater)? Questions that can be glossed over according to the vocabulary of modal hermeneutics in the following way: in what way is such and such a semantic aspect significant with regard to the esthetic problem that has arisen for the playwright, writer, sculptor, choreographer or composer? A question that finally hides another, which scientific prejudices concerning the "neutrality" of science have taught us to neglect, or even to consider as unwelcome – according to a point of view that regrettably confuses neutrality with a lack of critical self-reflection, taking advantage of a descriptive legitimacy *a priori* (logical, universalist, paradigmatic, "structuralist", etc.): in what way such and such an aspect is a *success* in relation to this esthetic problem?

There is a recursive effect here: the interpretation becomes interesting if it succeeds in highlighting how this or that aspect of a work is a success in terms of the general esthetic project and/or how the work as a whole is significant in terms of the corpus from which it emerges. The question is no longer that of a "true" interpretation, but that of a "significant" interpretation in the interpretative corpus field, in so far as it restores as yet unexplored aspects or renews the academic approach.

Significativity also makes it possible to integrate into the federative logic of modal hermeneutics all the non-verbal semiotic dimensions that a verbal work comprises, by recognizing the importance of the poly-semioticity of a text. For example, what is at stake in the visibility of Jacques Dupin's collection of poems entitled *Trait pour trait*? How is the organization of words on the page significant in terms of the overall esthetic project? *Significativity* seems to us to be a regime of meaning that must encompass the semantic dimension in order to be able to restore the singularity of the works by reconsidering the importance of their polysemioticity, as well as their pluri-parametric and pluri-normed character.

Michaux's poetry sometimes plays with the limits of verbal language and makes it tend toward forms of onomatopoeic associations that are at the limit of a-semantism. Faced with such verbal sequences, the following question arises once again: in what way can they retain their *importance,* i.e. in what way are they other than insignificant borborygmus, in what way can they remain culturally *significant?* To which other cultural traditions can they be related in order to assess the specific issues at stake in this work?

If it is not unworthy to consider – the most recent works in zoosemiotics lead us to think so, as the older works of Leroi-Gourhan do – that humans share certain communicative capacities with animals; the question of an "anthropological disruption" (Rastier 2018b, p. 25) thus remains central to the enhancement of this sphere of *significativity*, which highlights the capacity to consider cultural objects, here, works, as important, *significant*, in their links with *objects absent* from tradition, thereby underlining the exceptional importance of this semioanthropological dimension that François Rastier has called the *distal zone* (distinct from the identity and proximal zones):

> Even the theory of enaction, developed by Francisco Varela, a specialist in pigeon vision, maintains the continuum between animal and human. However, the neo-Darwinian evolutionary continuum neglects two-fold the semiotic *disruption*. (i) On the one hand, this disruption is manifested by access to the symbolic. Since the considerable semiotic activity of superior animals is limited to signals and clues, the symbol, in the Saussurian sense of the term, remains a semiotic of man's own, quite simply because symbols are organized in paradigms and thus require the manipulation of absent objects: if I say *black,* I exclude the other terms of the paradigm, first and foremost the *white* antonym. Not only distinct, but differential, symbols vary in meaning and expression depending on the context, which allows for the semantic opening of languages. The evocation of absent objects has as an anatomical substratum a developed prefrontal cortex, characteristic of man. It also allows the iconic creation, constitution and manipulation of mental images of absent objects. It opens up the past as well as the future, the categories of the possible and the counterfactual. (Rastier 2018b, p. 25, author's translation)

It is therefore a question of reconstructing, with the notional couple "stake"/"*significativity*", the possibility of thinking of another form of vital function, of a cultural type, by embodying and intensifying it as well; and by doing so, to reflect on the responsibility that the teachers have in the education of this type of semiotic life.

One cannot but be struck by the conclusion given by Jean-Michel Salanskis in Rastier et *al.* (1997):

> The inversion of the relationship between content and modality is what is required. From an ontological point of view, what counts is the content of the sentences and texts, because the content itself points towards the being, whereas according to a philosophy of meaning, the modality of delivery of the content is what prevails, because it is this modality that prefigures a reception, *it is the how of the content that governs making sense* [our use of italics]. (Salanskis 1997, p. 418, author's translation)

The mobilization of the notion of "making sense" corresponds in these remarks, it seems to us, to what we have referred to as "*significativity*". Jean-Michel Salanskis' critical reappropriation of such a philosophy of meaning goes far beyond the scope of our problem and diverts us from what we are saying: in a probably restrictive way, we wish to understand it here as a deontology – or practical philosophy – leading to an interpretative praxeology, which seems coherent with the rest of his remarks:

> Of course, this epistemological conception is part of a broader movement of the disavowal of ontology: the underlying philosophy is one that is not concerned with describing the constituent structure of being, but with deepening *our* understanding of meaning, what it means to live and think, particularly through theorizing and scientific knowledge. (Salanskis 1997, p. 416, author's translation)

Significativity, as a particular regime of functioning of understanding for a science of the arts understood as modal hermeneutics, thus puts the meaning of constituting a success (or a failure) and, by this, to *qualify* the meaning between us – but according to criteria that can be redefined each time according to the general aesthetic project and that therefore have nothing aprioric or prescriptive, as Dominique Rabaté emphasizes by insisting on the plural of the values potentially implemented:

> I prefer to talk about PLURAL values in art, according to a greater mobility of objectives. Leaving the dogmatism of the single or essentialist criterion, the passage to the plural also makes it possible *to consider the success* [our use of italics] of an art according to options or changing perspectives. (Rabaté 2007, p. 23, author's translation)

Significativity, which we propose to think of as the major regime of semioticity for music (of which non-ritual instrumental music is the most radically distinct example of verbal language), thus designates *the critical interplay that a work constructs with the cultural dimension*, including through verbal signified elements. By its mobilization, semiotics, conceived as material hermeneutics rewritten in a modal hermeneutics, could provide itself with the means to counter the accusation that a scientific approach to the arts would cause them to lose their intensity: in fact, its project is to restitute cultural objects conceived as factories of intensity, measuring their legitimacy against others through the elaboration of a singular *consistency*. Thus, the notion of *significativity* does not separate the question of intensity from the question of having a guarantee. The project of a science of the arts has an epistemological framework whose fundamental value is the preservation of the specificity of the objects – which ultimately, but fundamentally, makes it possible to link a semiotics of singularities to an ethics of plurality.

7.5. Modal hermeneutics and engagement strategy

In this framework, it is not therefore primarily *significativity for me* that is at stake, but the *significativity* of this or that aspect with regard to the general esthetic project – through which only the link between the singular and the collective, the meaning *for us,* is conceivable. Modal hermeneutics can thus compensate for a notable omission, that of an anthropology of literature (and of the arts), based on the question of action as an emancipatory effect, in short on the question of knowing "what literature *does to man*" (Rastier 2018a, author's translation). By adopting such a point of view, and by mobilizing the notions of "stake" and "*significativity*", the scientific project of the cultural sciences does not compromise intensity, and is even able to foster it. This is the pedagogical ideal on which modal hermeneutics leads: with a successful interpretation, a book, even if it is a *priori* disheartening because of its difficulty, always ends up pleasing because one has been able to reconstruct the stakes, stakes by which a work transforms a lifeless situation into a consistent, tense "world". The stakes that constituted the fundamental astonishment from which Cassirer thought about cultural objects: "How does the indeterminate arrive at determination, the formless at form?" (Cassirer 1945, author's translation)

As Dominique Rabaté reminded us in his introduction to *L'art et la question de la valeur*, it is now more than ever a question of describing and analyzing this "factory" of values that works of art represent. The urgency of reappropriating this question arises from the fact that it has for too long been dismissed by academics themselves, particularly in the name of the neutrality of science. In fact, if the Humanities were conceived as "soft" sciences, it can be considered that this is due, on the one hand, to a definition of scientificity that is biased because it is indexed on the model of the logico-formal sciences, and, on the other hand, due to the fact that the question of value has been expelled, even though it is the very place where the power of emancipation is affirmed through critical work: they would therefore be "soft" both through confusion and resignation. In fact, as Marie-Dominique Popelard and Denis Vernant, among many others, point out, the division between "hard sciences" and "soft sciences" is not insignificant, "which often leads to reserving credits for hard sciences whose technical repercussions are more immediate and visible".

Yet the authors do not fail to point out:

> the eminent political and social role that the human and social sciences should play. No natural science will, however, provide us with a means of managing economic crises and social conflicts, of avoiding wars and genocide. (Popelard and Vernant 1997, p. 84, author's translation)

Dominique Rabaté's refusal to ask the question of the value of works (which, in a certain way, also constitutes a refusal to submit oneself to evaluation) is situated in the 1970s, in a kind of compensation in the face of "the shaking of values induced by May 1968":

> We can also think that the shaking of values induced by May '68 was replaced in the 1970s by the triumph of the most formalist approaches to literature, approaches which have in common the refusal in principle to raise the question of the value of works. (Rabaté 2007, p. 11, author's translation)

With Blanchot, even if in a very different way, we find the same voluntary indifference to the question of value.

But is not it by confronting the question of value that the humanities (and more generally the cultural sciences, including linguistics) can claim any social utility by preserving the fundamental concern of "making science" and rejecting any tendency to conformism, which is sometimes allowed in "normal science" as Thomas Samuel Kuhn theorized in 1962 in *The Structure of Scientific Revolutions*?

As early as 10 years ago, Dominique Rabaté saw the "major lack of concern" of university critics with regard to the question of values as an "alarming symptom of resignation against which []... we must now react":

> The current situation of criticism and of the university seems to me more ambiguous: having assimilated all these approaches, with an ecumenism that is sometimes too benevolent, university criticism makes them function in a way that is free from the polemics or controversies that marked their emergence, by confronting them with largely canonical texts whose status is confirmed or renewed by its readings. Little involved in the artistic debates of its time, its task is therefore little evaluative – which is to be welcomed, but without this dimension being really assumed or claimed. At a time when the critical landscape is collapsing around it and promotion has increasingly blatantly replaced the critical act in the media, this major lack of concern seems to me today to be an alarming symptom of resignation against which I believe we must now react. (Rabaté 2007, p. 14, author's translation)

It is indeed through the question of value that we can first think of the "strength" of art and the "enchantment of the works that accompany and delight us" (Rabaté 2007, p. 6, author's translation).

In his review of Isabelle Kalinowski's book, *La science, profession et vocation. Suivi de Leçons wébériennes sur la science et la propagande* (Agone, 2005), Philippe Roman reminds us that the latter had the merit of giving all its depth to the question of *Wertfreiheit*, fraudulently translated by "axiological neutrality":

> First of all, it dispels the misconception that Weber was the champion of the scientist's non-engagement. He hesitated all his life between a political and a scientific career […]. In reality, as Isabelle Kalinowski points out, the interpretation of the concept of "Wertfreiheit" should not be built around the neutrality/engagement couple as Julien Freund did, but rather should be opposed to what Isabelle Kalinowski calls the "non-imposition of values". The problem of *Wertfreiheit* is therefore not that of the existence of values per se, or of adherence to values per se, but that of the dishonest use that can be made of values when they are present without being given as such by the teacher abusing the dominant position conferred on him by his position. (Roman, 2006, author's translation)

We find here the Cassirerian project of an antidogmatic cultural anthropology and the recognition by the semiotics of cultures of the need to adopt an "engagement strategy" (Gonseth), which is not an uncritical participation (or violent reaction) to the values constructed in a work as cultural studies can give examples, but on the contrary the project to restore their singular articulation, which makes their intensity and the success of their thyrse, to be able then, but only then, to take a stand, in a way that can be nuanced, complex, without radicalism or ideological bias.

Through the recognition of its status as modal hermeneutics, art history could find a way out of strict historicism to become transhistorical once again, by arousing the desire to understand the stakes of a work; an art science would find a way not to deprive itself of intensity, while demonstrating its critical capacity. The semiotics of *significativity* would reinforce the current specialists in ancient literature who are demanding the right to recognize the transhistorical dimension of their corpus[10]:

> Our attitude, our decision: we no longer have to conceive of classical writing as a form that must be defended as a past, legal, conforming, repressive form, etc., but on the contrary as a form that the rolling and reversal of history is making new [...]. In other words, we must today conceive of classical writing as being detached from the sustainable, in which it was embalmed. No longer caught up in the sustainable, it

10 We will recognize the position of a linguist like Claire Badiou-Montferran, whom I thank for this quotation, and who, for these questions, draws on the work of Hélène Merlin-Kajman and the 17–21 collective.

> becomes new; what is fragile is always new; we must work on it, this classical writing, in order to manifest the future that is within it. (Barthes 2003, pp. 373–374, author's translation)

If the decision is the same ("to manifest the future that is within it"), the semiotics of cultures allows such a project to be reintergrated within the framework of a comparative, materialistic and critical approach.

7.6. Science of the arts and the esthetic intention of the semiotics of cultures

By way of a conclusive broadening, we propose to shift the basis of reflection on the particular epistemological status of a science of the arts, by conceiving it no longer from the point of view of a typology of sciences, but from the point of view of the articulation of modes of thought, characteristic of the project of a hermeneutics of works – and perhaps of its specificity within the human sciences themselves, in the sense that hermeneutic praxeology would be a form of practical philosophy.

A science of the arts is in contact with what Simondon calls "esthetic thought", which would be, according to him, the commonplace, transductive, pre-existing to the dissociation between technical thought (as thought of the elements) and religious thought (as thought of the totality, independently of the history of beliefs):

> Aesthetic thought isn't simply a remnant [*souvenir*] of magical thought; it is what maintains the unity of thought's coming-into-being as it splits into technics and religions, because it is what continues to grasp a being in its unity, whereas technical thought grasps the being below the level of its unity, and religious thought above it. (Simondon 2017, p. 202)

This is why artistic objects have "the power to give rise to living gestures" (Simondon 2017, p. 203), a life-giving *action*:

> [...] Art is that through which a new reticulation emerges from out of science, morals, mysticism and ritual and as a consequence of this new reticulation, there is the emergence of a real universe, in which the effort, which had been separated from itself, and which arose from the internal disjunction that technics and religion underwent, comes to completion, and as a consequence of these two expressions of magic, the initial effort of the structuration of the universe. [...] Art aims at a

universe on the basis of human effort and reconstitutes a unity" (Simondon 2017, p. 204).

In doing so, by feedback from the object of study on the discourse that seeks to restore its singularity, a science of the arts would also have to deal with the problem of linking its technicality to extra-technical issues – a technicality for which it has nothing to be ashamed, but which it cannot consider as sufficient proof of scientificity if it is not linked to the explicitation of an epistemology, itself linked to extra-technical issues. In the context of a semiotics of cultures, these extra-technical issues can only be ethical, in the sense that the restitution of a singular universe is ultimately the testimony of an effort to encounter otherness in an adventurous way. In the sharing of this effort, the exploration for the human being of the possibility that an "esthetic judgment" – in the sense here of an evaluation of correctness (multiform, anti-dogmatic and anti-essentialist) – a relationship between the thought of the elements and the thought of the whole – can take place both in art and lives of those who experience it – of those who do it and redo it. The experience of translation is here exemplary of this ever-renewed chain of interpretation, in which the capacity to "mold" a world while constantly maintaining critical control over one's own *actions* is pushed to its most radical necessity.

In this way, a science of the arts conceived as modal *praxeology* would reduce the duplication of sciences in theoretical and practical mode, the consequence of which is the joint duplication of science and ethics.

By providing access to the worlds of works of art, it would promote the "real encounter between different modal orders of thought" (Simondon 2017, p. 207), giving "real and ultimate satisfaction considered as vitally experienced" (Simondon 2017, p. 207), while "making human achievement unfinished" (Simondon 2012, p. 274), salutarily distinct from adherence to "the mythology of a group [...] erected as a universalizable doctrine" (Simondon 2017, p. 233).

A science of the arts would thus constitute a fruitful solution to the "insoluble aspect of this problem of the relation between science and ethics" in the sense that it would achieve, in an *interpretative action*, a coherent and unified synthesis of theory and practice, by which the "encounter of technicality and respect with regard to totality" would not be reduced to a banal "alliance of an ensemble of procedures and of a mythology" (Simondon 2017, p. 233) – thus revealing its *significativity* for the definition of culture as a possible place of "de-excysting of prejudices" (Dessons 2004) and the ferment of a paradoxical fraternity, because it is based on the unexpected – where what we find is not always what we seek. Modal hermeneutics, centered on the arts, would allow us to explore the hypothesis that humans are defined at least as much by verbal language as by their ability to handle, in verbal language and in other non-verbal semioses, *significativity* in emancipatory terms.

Significativity thus sheds new light on the Cassirerian position, which was recalled at the beginning, that art cannot be taken further without a linguistic approach, and that language cannot be understood in all its depth without an artistic approach to culture.

7.7. References

Adorno, T.W. (1963). *Der getreue Korrepetitor. Lehrschriften zur musikalischen Praxis.* S. Fischer, Frankfurt.

Barthes, R. (2003). *La Préparation du Roman I et II. Cours et séminaires au Collège de France (1978-1979 et 1979-1980).* Le Seuil, Paris.

Borio, G. (1996). Analyse musicale et herméneutique. À propos de la reconstitution du sens en musicologie. In *La musique depuis 1945. Matériau, esthétique, perception*, Dufourt, H., Fauquet, J.-M. (eds). Mardaga, Liège, 81–94.

Bouquet, S. (1997). *Introduction à la lecture de Saussure.* Payot, Paris.

Dalhaus, C (1990). Analyse und Werturteil. In *Musikpädagogik und Lehre*, vol. 8. Schott, Mainz.

Dessons, G. (2004). *L'art et la manière.* Champion, Paris.

Dondero, M.-G. (2019). *Les langages de l'image. De la peinture aux Big Visual Data.* Hermann, Paris.

Goodman, N. (2006). *Manières de faire des mondes.* Gallimard, Paris.

Hakim, N., Dufourcet, M.-B. (1991). *Guide pratique d'analyse musicale. Cours, Lexique illustré, Tableaux.* Combre, Paris.

Hatten, R.S. (2002). Théorie de la musique et sémiotique générale: une interaction créative. In *Questions de sémiotique*, Hénault, A. (ed.). PUF, Paris, 565–586.

Kuhn, T.S. (2008). *La structure des révolutions scientifiques.* Flammarion, Paris.

Ladrière, J. (2005). *Les enjeux de la rationalité. Le défi de la science et de la technologie aux cultures.* Liber, Paris.

Latour, B. (2012). *Enquête sur les modes d'existence. Une anthropologie des Modernes.* La Découverte, Paris.

Meyer, L.B. (2011). *Émotion et signification en musique.* Actes Sud, Paris.

Molino, J. (2009). *Le singe musicien. Sémiologie et anthropologie de la musique, précédé de Introduction à l'œuvre musicologique de Jean Molino par Jean-Jacques Nattiez.* Actes Sud/INA, Arles.

Nattiez, J.-J. (1975). *Fondements d'une sémiologie de la musique.* UGE, Paris.

Nattiez, J.-J. (n.d.). Analyse et sémiologie musicales [Online]. *Encyclopaedia Universalis*. Available at: http://www.universalis.fr/encyclopedie/analyse-et-semiologie-musicales/ [Accessed 22 August 2019].

Popelard, M.-D., Vernant, D. (1997). *Les grands courants de la philosophie des sciences*. Le Seuil, Paris.

Rabaté, D. (ed.) (2007). *Modernités n°25. L'art et la question de la valeur*. Presses Universitaires de Bordeaux, Bordeaux.

Rastier, F. (1987). *Sémantique interprétative*. PUF, Paris.

Rastier, F. (2011). *Arts et sciences du texte*. PUF, Paris.

Rastier, F. (2016). *Créer: Image, Langage, Virtuel*. Casimiro, Paris-Madrid.

Rastier, F. (2018a). Poésie et langue. Aspects théoriques et didactiques [Online]. Available at: http://journals.openedition.org/pratiques/4907, January 2019.

Rastier, F. (2018b). *Faire sens. De la cognition à la culture*. Classiques Garnier, Paris.

Rastier, F. (2018c). Cassirer et la création du structuralisme. *Texto ! Textes et cultures*, XXII(4), 1–24.

Rastier, F. (2019). Sur la sémiotique des cultures. Entretien entre Lia Kurts et François Rastier [Online]. *Fabula*. Available at: http://www.fabula.org/atelier.php?Linguistique_et_semiotique_des_cultures [Accessed 20 August 2019].

Rastier, F., Salanskis, J.-M., Scheps, R. (eds) (1997). *Herméneutique: textes, sciences*. PUF, Paris.

Roman, P. (2006). Max Weber, Isabelle Kalinowski, La science, profession et vocation. Suivi de "Leçons wébériennes sur la science et la propagande" [Online]. *Lectures*. Available at: http://journals.openedition.org/lectures/303, February 2019.

Simondon, G. (2012). *Du mode d'existence des objets techniques*. Aubier, Paris.

Vaugeois, D. (2006). *Revue des Sciences Humaines, 283. La Valeur*. Presses universitaires du Septentrion, Villeneuve d'Ascq.

van Vliet, M. (2016). Art et langage chez Ernst Cassirer: morphologie et/ou structuralisme? *Images*, 5, 1–22

8

The Roles of a Semiotics of the Arts: Working Hypotheses for Overcoming the Shortcomings of the Past

8.1. Some remedies for previous theoretical abuses

8.1.1. *Partial approaches, all powerful*

The study of works of art has historically unfolded all its complexity, which is paradoxically reflected in both the reductionism of approaches and the use of a thaumaturgic idealism. Indeed, on the one hand, a specific aspect of the work has been emphasized while forgetting the others: (1) its independent, even *autotelic* nature, homologated on the basis of the model of the allographic arts, and therefore of a notational score dissociated from the historical context of production; (2) its establishment on the basis of a *techné* and an executive performance, which links it indissolubly to a biography and a sociocultural environment; (3) its expressive capacity to shape what was not yet *one* on the cognitive and emotional level; this would guarantee art an unrivalled communicational potential.

On the other hand, a kind of *Aufhebung* (therefore conciliation and elevation) of its multiple and contradictory aspects was carried out in order to exalt alleged contents of truth: here is then the opening of the worlds capable of rooting values of the absolute, the guaranteed manifestation of beauty or harmony. To this can be added the most formal version of idealism, namely the *institutional theory of art*. Through the observation that an object's access to artistic status is a performative act in the art world, this theory seems to render the analysis of the work and even its

Chapter written by Pierluigi BASSO-FOSSALI

actual subsistence almost useless, in favor of a mapping of poetics that manages to be validated and admitted to the supermarket of contemporary aesthetics.

While the *historical-institutional theory of art* (Levinson 1979) seeks to safeguard at least the historical genealogies of the "artistic proposition and social validation" couple according to the reconstruction of intentional and conventional frameworks of relevance, in the most nominalist postmodernist version, the critical aspect is totally lost.

As for semiotics, it has often risked mixing its radical textualism[1] with the idealistic prejudice of the *autotelism* of the work of art and reducing the aesthetic *quid* to a specific linguistic functioning: for example, the projection of the *principle of equivalence* from the paradigm plane (axis of selection) to the syntagmatic plane (axis of combination). *Self-reflexivity* was then conceived as the internal play of art without the need for any hetero-referential scope, down to a formalistic reduction (appreciation of the *plastic* composition as such, possibly correlated to very general semantic components, of a mythical or cosmological order).

This reductionist and partly idealistic, even romantic, vision of the (poetic/artistic) work has not failed to leave us interesting results and still significant theoretical directions, such as the fact of grasping in the Jakobsonian principle of projection a search for analogies and homologies internal to the work, according to a desperate or joyfully unsuccessful attempt to remotivate the arbitrariness that intertwines the plane of expression and the plane of content[2].

---

1 Semiotic analysis or scrutiny – to stick to Wollheim's criticisms from his analytical aesthetic approach – must also pass through an circumstantial investigation and an exploitation of personal collateral knowledge, so as to understand the discursive "gestures", and not only the internal weft of the signs (text). For example, it should be known that Sassetta, in the realization of the *Dossale di San Francesco e sei miracoli*, used the most expensive and difficult pigment available to paint the tunic that the saint gives to the poor knight (Wollheim 1980, p. 129). But what collateral knowledge is relevant? Can knowledge of the context of production distinguish institutions of meaning from non-determining conjunctural connections, the roots of the creative act from anecdotes? Wollheim (1980) puts forward doubts about a criterion that could be legitimate: "Retrieval is legitimate because, but only in so far as, through its findings it contributes to perception" (p. 131). The criterion seems to impose constraints, such as the perceptibility of clues and the connection with the effectiveness of the aesthetic experience. However, the misattribution of a painting can be a demonstration that, once the framework of knowledge has been corrected, perception cannot be changed; what changes is a belief system around the painting.

2 See the theory of semi-symbolism (Greimas 1989). For a critical account of this notion, see Basso Fossali (2018).

As for Nelson Goodman's theory, with its distinction between autography and allography and its "symptoms of aesthetics", it could only belatedly meet with interest and development in semiotics, in view of a critical reexamination of the identification between text and work (Genette 1997) and a broader spectrum of specific semiotic functions. At the same time, with the concept of *implementation,* Goodman (1984) has both supported and exercised an implicit critique of the "institutional theory of art" that is incapable of prolonging reflection on the forms of meaning of the work when it is inserted in a specific space of aesthetic experience.

It was not until the 2000s that semiotics finally saw a systematic shift in semiotics from the isolated study of the work to its insertion within a corpus, even if diachronic studies were still very little developed. Thus, although art has been the preferred field of application of the semiotic method, there are still remarkable delays.

The semiotics of the arts to come will have to recognize its implications in the reductionisms and idealisms of yesteryear without making the mistake of repudiating its past and globally delegitimizing the currents of contemporary aesthetics. On the contrary, even an author like Collingwood, with his *Principles of Arts* (1934), beyond aesthetic idealism, has shown how to approach art according to an articulation of method and discovery that would be internal to his practices. Today, we could link this articulation to a theory of art in which one would postulate either a refusal of any method *a priori* (free invention) in order to make semiotic organizations emerge afterwards, or a scrupulous process in order to meet the inflections of chance or the unexpected answers of the matter. This dialectic is an implicit criticism of a demiurgic vision of art in order to inscribe it in a paradigm of semiotic ecology.

8.1.2. *A hermeneutic paradigm for a semiotic ecology*

The link between semiotics and the arts could only be consubstantial to the common defense of a cause: the capacity of verbal and non-verbal languages to offer a meaning of their own, but autonomously and differently. At the same time, they enter into a productive relationship of reciprocal translation that cultivates meaning but also senses. Yet, in the eyes of other human sciences, an attempt by the semiotics to legitimize the ambitions – even theoretical and philosophical – of non-verbal languages has turned into a pretention to reduce their expressions to sets of figures and signs, to detect codes in their textual repertoires, to guarantee an unambiguous interpretation of the works. Thus, even today, we wish to underline the resistance of the arts to the "clinical" view of a semiotic science that is incapable of dealing with matter and expressive media (for a discussion of Elkins' 2017 theses, see Basso Fossali (2018)) as well as with the spectator's embodied perception because of an overly cumbersome legacy from the theory of the linguistic sign.

Undoubtedly, semiotics was tempted to make its procedures easier and to operate a series of reductionist hypotheses, but this approach was counterbalanced by the complexity of the modeling already required for analyses with reduced or even basic "granularity"[3]. Although today's semiotics is no longer linked to the reductionist, albeit operative, visions of the 1960s, aesthetics, with a few exceptions, has increasingly distanced itself from a discipline that is still discredited as "sign and code sciences".

In other contributions, we have tried to thematize the reasons for the misunderstanding between semiotic and aesthetic theories and the potentialities of their possible dialogues (Basso Fossali 2002, 2011). Beyond the observation of the uselessness of persevering in the same critical path, it is highly probable that the future perspectives of a semiotics of the arts are no longer linked solely to the confrontation with the proposals of analytic aesthetics or neuroesthetics. Keeping up with current events, or even intellectual fashions, implies the risk of forgetting meritorious traditions of thought and the always opportune dialogical complementarity between different theoretical perspectives in the face of the complexity of the corpus of study.

The need to go against the tide can be embodied today by the defense of a hermeneutic paradigm banished by chronic consumption patterns or by a manipulation of meaning *ad libitum* in episodic consumption that no longer needs the critical devotion of interpretation. A phenomenology of standardized receptions can do without the attestation of privileged encounters between *works* and *interpreters* and without attesting to the catalysis of new meanings in the coupling that binds them. This is why meaning is, on the one hand, grasped as an activation and expenditure of values, according to an economic reason of corporality and society, on the other hand, objectified as an attribution of semantic potential according to a statistical approach[4].

It is clear that considering meaning as a constantly renewing issue that requires a diagnostic evaluation of its finalizations risks no longer being accepted either as a scientific task or as an ethical requirement. In spite of everything, a semiotic ecology of culture remains a project that is still poorly digested and widely depreciated because it does not guarantee immediately homogeneous and repeatable results;

3 This metaphor comes from the notion of "photographic granularity" and indicates here the gradual passages between microanalysis and macroanalysis.

4 Umberto Eco has always stressed the opposition between an interpretative meaning and a meaning based on a "statistical extrapolation" of attested linguistic usages (see Eco (1968)).

moreover, it requires a strong problematization of the relationships between the social actors concerned, their fields of practice and their environment.

Semiotic ecology is rooted in a rather basic aspect of experience: perception, act of utterance[5] and interpretation can only select a limited number of values (economy) in relation to the "density" of (i) the sensible world, (ii) the social-historical framework and (iii) the textual space. However, the intended meaning must also face the totality of the confrontation with the environment that includes it (dependence extends the capture of values). Ecology shows a kind of *double bind* of signification: at the same time, it is necessary to choose, that is to say, to make selections in order to "make a difference", and to accept being overwhelmed by the complexity of the environment, which qualifies the coupling between the subject and his *milieu*. On the semiotic level, ecology shows that, despite the limitations of relevance of our points of view, indeterminacy or the "not-yet-relevant" haunt our choices (spectral dimension of discourse, see Basso Fossali (2017, section 6.4)).

Once it has been decided to attribute a *form of life* to the subjects – a form of life that cannot be superimposed on identities forged and exchanged through language games – the same must be done with cultural objects. In this regard, Rastier emphasized that "the artwork never discovers its finality, and even when completed, remains unfinished, inasmuch as it can be indefinitely reworked by other artworks" (Rastier 2016, p. 27, author's translation). An immediate consequence of this idea is that a nominalist attribution of *artisticity* (Eco 1989, p. 172) to an object cannot resolve the questioning of its form of cultural life.

8.1.3. *Skepticism and responsive aptitude*

In relation to the exercise of performative statements that can establish, by a kind of "blessing", the artistic object as such, we can assume a skeptical position. Stanley Cavell, in his book on Shakespeare *Disowning Knowledge*, has shown that skepticism is above all: (i) a distancing from what one possesses; (ii) a critique of knowledge, the latter being constructed normally under optimal conditions of experience. At the same time, skepticism does not accept the separation of otherness, just as it does not accept the construction of a transcendent horizon that is distant and intangible, even virgin. Even beyond the relevance of his interpretation

5 "Utterance" cannot accurately translate the French term "*énonciation*" because the latter refers to a theory of discursive production that concerns all semiotic forms. On the one hand, it should be pointed out that alongside "speech acts" there are also "pictorial, musical, etc. acts"; on the other hand, the "semiotic enactment" of the potentialities of different languages builds a possible world of discursive values within which the discursive instance must take positions (points of view) and forms of involvement (modalities).

of Shakespearean skepticism, Cavell proposes to recognize paradoxically combined aptitudes in art: on the one hand, its *responsive* and *emancipatory* character; on the other hand, its propensity to recognize a *shared involvement* and constant *interpenetration* between identity and otherness. This double bind only reveals a critical, uncomfortable, even paradoxical position of the discursive and interpretative instances mediated by the work of art. One can immediately appreciate the contrast with the so-called *aesthetic distance*.

Cavell underlines the *mise en abyme* of skepticism: from the *possible world* of Shakespearean characters, he returns to the instance of discourse and vice versa. This should guarantee the *positional power* of art, which translates above all into an uncomfortable invitation to awaken perceptive potentialities and to mobilize as responding subjects. On this point, we can find full convergence with Cassirer's lesson: if theory demands evaluation and possibly adequacy (conformity of use), art demands response, a "delayed" response (Cassirer 1944, p. 43). The interpellative power of art is a driving force of history, a history made of reappropriations[6].

Reappropriation – found in Wollheim's aesthetics (1980) under the concept of *retrieval* – does not concern the reappropriation of the artist's intentionality as such, but the vocation to play the role of wise or skilful mediator between institutions of meaning and the artwork. Of course, there is no need to favor the cognitive over the emotional or the opposite: the aesthetic *trait* proves to be frequently concerned with the attempt to ward off any somnambulistic or apathetic state in the face of historical drama.

Art is not entrusted to shallow aestheticism if its attachment to *in vivo* experience means the actuality of the *link* with a present that is historicized because of the artwork; and in this link, we should not see the transparency of a truth, but the need to reconstruct a double bind with the present: to be *responsive*, by taking

6 Cassirer emphasizes more the individual aspect of the experience, even if as an opening of the subject to the multiplication of their living possibilities (Lauschke 2010, p. 41). It is on the basis of this clarification that Cassirer's idea that "art relates to 'the inner reality of the life process'" (Cassirer 2004, p. 387, quoted by Lauschke 2010, p. 34, author's translation) must be re-read and corrected, and thus "The work always remains – insofar as it stands purely on its own – simultaneously the testimony of an individual form of life, an individual Dasein and a particular kind of being" (Cassirer 2012, p. 46). Subjectivity is evoked as a support for a problematization of meaning and not at all as a horizon for a solution. In this sense, interpretative reappropriation cannot concern the artist's original intentionality; the artwork's appeal is to take on a gnoseological mandate for possible discoveries that it only intends to mediate. The timelessness of an original meaning is thus contested as well as a reappropriation as a re-actualization of the meaning of the work (because of the interpreter it would speak in the present tense of the "present"!).

advantage of a spatio-temporal distance, and to be *involved*, by refusing to believe oneself separated.

In the approach to interpretation, the very attitude of skepticism reveals itself as a willingness to recognize one's own aporias and blindness: more generally, a proactive skepticism seeks to grasp the patterns of social meaning so as not to learn it too soon or too late – as the figure of Macbeth, studied by Cavell, shows very well. Prophecy as well as *ex post* commentary and announcements as well as farewells must be discarded, accepting an *ecology of non-knowledge* (Luhmann). Thus, in order to be an interpreter of the present, art must most often accept to be untimely, to present itself as a work out of play in relation to the apparent self-understanding of social matters.

8.2. Some remedies for the universalization brought about by postmodernism

8.2.1. *A non-ethnocentric aesthetics*

A welcome *reentry* of the skepticism of about our argument is the recognition that a general idea of art cannot be abusively imposed. Any working hypothesis in the field of art should accept to go through a critique of this generalization, and if it is finally possible on some levels, it is because its validation has passed through comparative studies concerning other cultures and times. Clearly, this task goes beyond the limits of this contribution and of our competences as well, but we do not want to give up proposing a promising path of questioning.

Despite the strong ethnocentrism and idealism that characterizes thinking about art, the comparative and skeptical approach to our certainties is not new. One example, which is both revealing and memorable, can be found through the figure of Ernst Grosse, which we will use to begin our investigation. He worked in Freiburg im Breisgau toward the end of the 19th Century on a highly articulate and ambitious project to establish an ethnological aesthetic. In his book *Die Anfänge der Kunst*, published in 1894 and translated into English in 1897 under the title *The Beginnings of Art*, Grosse presented the following theories:

a) there is no people without art;

b) the science of art must extend its studies to all populations;

c) all civilizations and all forms of art have an equal right to science;

d) any work of art taken in itself is only a fragment. To be complete, the artist's work needs the spectator's ideas; only in this way is born everything that the artist wanted to create;

e) it is impossible to excuse a scholar who in our time is constructing theories about art without knowing that European art is not the only art that exists, art in itself;

f) we have never even been bothered to ask the natives for the signification of their drawings[7].

Reading Grosse's book, beyond certain initial biases for functionalism (the social utility of art) and for *mimesis* (the imitation of nature guides all works), one is faced with a very polemical intellectual project against ethnocentrism. This project is : (i) open to all the arts (for Grosse, ballet or popular music are part of artistic expression); (ii) characterized by scientific pretensions that require a comparative approach according to a balance between – we would say today – an *etic* approach and an *emic* approach; (iii) animated by an intersubjective vision of artistic creation that defends the theory that the life form of the works is established and can evolve through its spectators or readers.

In 1927, Boas (1955, p. 14) presents the work of Ernst Grosse, emphasizing that Grosse highlighted the articulation in art between practical purpose and aesthetic function, the latter not being ornamental. On the contrary, even "primitive ornament is by origin and by its fundamental nature not intended as decorative but a practically significant mark or symbol". Boas considers that Grosse indicates a "practical significance [implying] some kind of meaning inherent in the form" (Boas 1955, p. 15). However, it must be understood that this form is an exercise of the nature (a vital attitude) or of the technique (an instrumented behavior), in short, a significant expression of a search for finality.

If welcoming the "art of earliest times" into aesthetics means finally acknowledging that it "is at one with the art of all times" (Grosse 1897, p. 307), then the autonomous and purposeless character of Romantic art is only an ideology that has, in reality, had its techniques (rhetorical as well) and its forms of participation in a more global semiotic ecology. Comparatism must integrate difference and try to understand the role of art in the evolution of civilizations and in the sensitive forms of socialization of material and intangible values.

For Grosse, art has two characteristic modalities: the ability to activate an "immediate emotional factor" (Grosse 1897, p. 48) and the ability to translate other social issues (politics, religion, economics, etc.). Added to this is the conviction that art is not the latest elaboration of an emancipated culture, but the strongest contact

7 In fact, the six theories are quotations from Grosse's book, extracted and presented by the publisher at the beginning of the new 2009 edition in French. They are less starting points than conclusions. For example, the first theory – "there is no people without art" – is stated at the end of the book (Grosse 1897, p. 312, 2009, p. 315).

with nature: abstraction is also *mimesis* of nature, to extract forms as a learning process about the stability and complexity of organizations, and to develop techniques that can be used elsewhere (Grosse 1897, pp. 311–12; 2009, p. 315). But the most remarkable consequence of art's closeness to nature is that in its practice, even if it is devoted to elevating minds, there is still a trace of its original vocation: to participate in the struggle for existence, a resistance (Grosse 1897, p. 314).

Beyond the evolutionary ideology that inscribed Grosse's thought in his time and which was the element strongly criticized in 1928 by Moritz Geiger (Geiger 2005, p. 189), there is the idea that the search for aesthetic and technical forms was merely a framework of convergent, interdependent values. This is Grosse's (1897, pp. 311–312) fundamental argument against Herder and Taine for rejecting the determinism of biological and climatic factors on the evolution of culture. Art is a production capable of its own evolution and a partially emancipated response to external factors, in short, capable of distinguishing itself from natural evolution. Art would then be involved in a kind of negative evolutionism, a programmatic and punctual deviation of culture from any form of pre-organization, even if it is natural or linked to an autonomous social domain. Art seems to generate a new symbolization waiting for an answer, because the work is only a fragment that calls for the integration of a new design. For this reason, foreign art can only enter aesthetics as a squared fragment, both autochthonous and cross-cultural.

8.2.2. *The (un)manageable nature of primitive art*

While the ethnological aesthetics proposed by Grosse is sometimes marked by an excess of generalization and by some questionable preconceptions, the debate on the primitive art that succeeded him has often shown a discreetly reductionist vocation by dealing, on the one hand, with African masks in Picasso's work and, on the other hand, with the dubious character of the nominal label "art" applied to non-Western aesthetic productions. The discussion on the content of legitimate practices – through a scholarly makeshift job one can reframe the "primitive" – and on the social functions of Western art has given as a paradoxical result the nominalist drift of art, any object being able to follow a protocol of candidacy to aesthetic appreciation. And even this candidate-object was soon dismissed as a now useless prejudice in view of the legitimization of conceptual or relational art.

In her book *Primitive Art in Civilized Places* (1989), Sally Price attempted to present the art of the Maroons of French Guiana and Suriname, based on the reconstruction of the "legitimate aesthetic framework", i.e. with "its own history". But this philological attempt implied or induced rejection: (i) by Western art museums, because "any ethnographic contextualization" was considered "alien to the aesthetic character of the objects"; (ii) by ethnographic museums, because classifying objects on the basis of the nominal label "art" was an ethnocentric abuse (Price 1989, pp. 89–122).

However, primitive art was able to serve, on the one hand, to claim an instinctive sensibility below cultural knowledge[8] and thus to exemplify a liberation of the eye of the connoisseur as an "organ of tradition"[9]; on the other hand, to note the lability of cultural status and functions of art, to such an extent that primitive art is easily reduced to an "exotic curiosity". The latter is, moreover, the only possible valorization in the face of the loss of the object's genetic context or its postcolonial contamination. Moreover, if the original implementation framework of the primitive work is reconstructed, it is then perceived as a useless crutch that shows in itself the inadequacy of the object in relation to the concept of *art*.

Fortunately, after Grosse and Boas, an ethnological aesthetic has developed that has not failed to give some good scientific contributions[10]. That said, a whole series of questions have not been satisfactorily answered. Can comparatism escape the project of finding lines of convergence, some properties shared by all aesthetic productions, or even a specific status for art within the semiotic ecology specific to each culture? Is art a symptom of an institutional paradox that is part of the history of each culture? Is there a structural nature of art that can be studied, without claiming to generalize the narratives of legitimation and the practices of enjoyment and evaluation that each culture seems to shape throughout its history? Is the cosmopolitanism of art a federative project of distancing the institutions of meaning internal to each culture?

In relation to these questions, the contribution of semiotics is still of the order of *desiderata*: at the same time, we would like to find a specific and stable symbolic role for art, but we can only note the gaps that prevent a positive, articulated and reliable answer. As far as the anthropology of art is concerned, the teachings of semiotics are despised and thus non-verbal languages still need tutoring. In this regard, in the catalogue *La fabrique des images*, Descola argues that "images do not speak for themselves" (Descola 2010, p. 11, author's translation), which is why their readability is usually entrusted to "conventional categories that are purely descriptive" (Descola 2010, p. 11) of a historical or geographical nature. The risk of reproducing the defects of an iconography that projected literary content onto images in such a way that the visible was merely an illustration with at most a few variants. Having said that, the subjection of the image to literature can be dispensed with by attributing to it a revealing power.

8 Artistic instinct is also at the heart of Grosse's (2009, p. 311) argument.

9 Sentence attributed to Franz Boas (Price 1989, p. 22).

10 Looking back over the last few decades, we can mention, among others, the work of Hans Belting, Philippe Descola and the journal *Res*.

Descola's idea is that the task of contemporary anthropology is to reconstruct the ontologies underlying cultural practices in such a way as to make explicit the true perception of reality on the part of a given community, a perception that is always filtered through institutions of meaning. In his major scientific contribution *Beyond Nature and Culture* (2013b), Descola notes that what changes between naturalism, animism, analogism and totemism is the *inferential* regime of meaning, as there are different presuppositions in the constitution of a world of reference. This idea is applied to the world of art and in particular to the meaning of images, as shown in the exhibition *La fabrique des images,* held at the Musée du Quai Branly:

> The idea that structures this exhibition is that identity judgements are not only expressed in statements, that they are visible in images, that to depict, in short, is to show the ontological framework of reality. (Descola 2010, p. 17, author's translation)

Cultural type is predictive of a representational and intentional regime, which gives us access to the "system of qualities expressed in images" (Descola 2010, p. 17, author's translation). Yet each image has the capacity to interpret this value system in the sense that it allows it to manifest itself more or less effectively.

> Thus understood, figuration does not refer to formal recipes, iconological archetypes or procedural systems; it makes manifest the links between the structure of an ontology and the means used to make an image represent this structure and make it active. In short, it is a morphology of relationships that we are trying to highlight here, not a typology of forms. (Descola 2010, p. 17, author's translation)

Beyond this *cultural enaction* of the ontological framework promoted by images, Descola (2010) admits that there are "expressive functions, if not absolutely universal, at least extremely common" (p. 17, author's translation). These functions are those relating to pictograms and coats of arms, where there is a strong codification. Apparently, it is the rate of conventionality that reduces ontological differentiation and systemic divergence between one culture and another, which could be counter-intuitive.

The explanation is as follows: the image that does not respond to strong codes is not "mediated by a discourse", because it is "directly visible in form and content". Thus, visual art can exploit its "active causality", its effectiveness being the recognition of an "echo of the ontology with which it is familiar" (Descola 2010, p. 18, author's translation).

It is clear that primitive or contemporary works are merely multiple reflections (echoes) of an underlying ontological structure without linguistic mediation and endowed with directly accessible signifying power.

Despite this epistemological vision of the meaning of images, which is apparently contrary to a semiotics conceived as a science of mediations, Descola's lesson had a remarkable impact on this disciplinary tradition. By discussing the specific roles of "culture" and the semiosphere in our civilization, given their opposition to "nature", we want to show that there are other possible conceptions, well attested in other civilizations. While this objective is laudable, although not new, we should question the status of culture in other civilizations or the very abuse of this category (which reproduces the question of primitive art on another scale); then we should clarify whether we intend to recognize the weaknesses of the institutions of meaning in other civilizations, or whether we prefer to think of a monolithic ontological architecture that frames any production of meaning.

8.2.3. *In search of a meaningful place*

The concept of art remains rather implicit in this debate on the manifestation and evocation of the ontology underlying the culture of reference. In any case, it should at least be understood whether art is simply interpreted by ontological regimes or whether, on the other hand, it actively participates in their elaboration and renewal.

In the past, with a classic formula of contemporary anthropology, art has already been attributed a very general function, that of *displaying the symbolic structure* of a civilization. But by stressing that art imposes itself as a "deep play" – to use a typical formulation of Clifford Geertz (1973) – one can only run the risk of acknowledging that this general function (displaying the symbolic structure) is also exercised by practices that have no aesthetic vocation.

Moreover, in the eyes of Descola (2013a, p. 36), Geertz is only the "talented advocate of hermeneutic anthropology" who nevertheless made the mistake of talking about an adaptation of man to his environment, thus reifying the presence of "nature". In reality, this defender of interpretative anthropology has upheld a vision of man as an "unfinished animal" (Geertz 1973, p. 49) that finds in the inextricable relationship between culture and nature its only condition for survival.

Specifically regarding the subject of our investigation, Geertz has clearly denied that art is a modality of exhibition and consolidation of social values. In this regard, he used passages from an article by Goldwater[11] (1973) to argue that artistic texts are "primary documents; not illustrations of conceptions already in force, but conceptions themselves that seek – or for which people seek – a meaningful place in a repertoire of other documents, equally primary" (Goldwater 1973, p. 10 quoted by Geertz 1983, pp. 99–100).

As an aid to research rather than stabilization, art plays an equal role with other social domains, but enhanced where there is no institutional sense that saves the *third-party role* and the illusory founding character that they would like to embody. The "primary" character belongs to an *instituting institution* always in paradoxical search for what it has actually instituted, exactly as in the case of an invention in the face of its future *symbolic effectiveness*, which is still imponderable (even the law has its unpredictable perlocutionary effects).

Another important element in Geertz's work is that art is conceived as an articulation between complementary and "appropriate skills" (Geertz 1983, p. 103), and in particular as a meeting between technical and analytical capacity[12]. This may help to clarify the relationship between the diversity of "primary" documents in search of their "meaningful place" (Goldwater 1973). By operating both a synthesis (technical integration of diverse materials) and an analysis (critical discrimination of normally joint values), the arts signal the path of a possible redevelopment of a cultural landscape through a transfer of relationships, valences, modal loads. But if there is only a "passing of the baton" in a coupling between works and interpreters, both in search of their renewed place, it is because, in general, any totalization of a landscape of meaning is impossible[13].

There is a real appropriation of the work, against any act of consumption, when the recognition of a dose of fallibility in the interpretation is articulated with the aim of the interpreter to characterize the object analyzed as accurately as possible, according to a competence that remains to be built and a new place to be attributed to it, which the work redefines each time. As for the analyzed object, the work alone can guarantee the balance between the two sides of appropriation ("appropriating" and "being proper to"), but only on the condition of escaping from what the spectator already knows (Geertz 1986, p. 132, quoting Baxandall 1972, p 48: "The

11 Founder and first director of the Museum of Primitive Art in New York and husband of the artist Louise Bourgeois.

12 This is a point of contact with Baxandall (1972) and – we would say – also with Grosse's (1894) vision.

13 As soon as there is a historiography, it can only continue to point out the impossibility of a totalization of meaning (Koselleck 2011, pp. 138–39). Moreover, even "the tongue is a dress made of patches" (Saussure 1907, p. 132, author's translation). Can we conceive of ontologies as architectures without ruptures or internal paradoxes?

public does not need [...] what it has already got"). Moreover, looking inside the object under analysis does not mean participating in a space of homogeneous and self-explanatory values, but finding a *distance* between primary sources that are not yet reconciled. In art, *techné* is not satisfied with itself but with its capacity to problematize the conditions of apprehension and conceptualization of a cultural component, to act as a catalyst for an aesthetic, affective and epistemic movement.

What the "period eye" (of the *Quattrocento*, according to Baxandall's 1972 study) teaches us is to make oneself available to look at what shows a negative posture, a resistance, a fracture that stimulates participation in other forms and measures. This is a good explanation of a paradigm of *complexity* in the approach to art. Our view of the otherness of an artistic era cannot be satisfied by reconstructing the competence and ontology of reference, because the recognition or the correct inscription in the institutions of meaning was not the specific goal of the experience mediated by art.

Beyond the various manifestations and civilizations, one may ask whether the structural role of art is merely that of displaying the emancipatory capacities of a culture in relation to its institutions, which are normally very concerned with having to show sooner or later their secondary, even internal, self-referential facade. But what kind of emancipation would art be the manifestation of?

In our opinion, the crucial problem of art cannot be limited to figuratively enacting the ontological regimes of an *épistémè* and thus accompanying a regulated distribution of the *phusis* and *psyche*. Art seems to be exploring semiotic regimes used by practices whose legitimacy should never be taken for granted, even if they self-affirm the meaningfulness of their operations. In this sense, art wants to question the privileged relationship between linguistic mediations and institutions, a relationship that functions as the cornerstone of *symbolization* systems. Actually, art is not an epistemology in the strict sense of the term, but would nevertheless indicate the possibility of a different life of signs, an alternative ecology in which any rigid (coded) symbolization cannot claim to frame and saturate the meaning of *forms of life* through fixed, ontologically determined identities. The risky ambition to generalize the role of art in relation to cultural differences would then be partially motivated by the fact that art precisely proceeds to an erosion of founding ambitions of codes, with deviations, tensions, subversions of the administered or attributed meaning.

Thus, access to art could be grasped as a kind of anti-institutional pedagogy, as it is the constant search for meaning that explores an innovative exemplification, not its binding stabilization. At the same time, through art, we show the primary dignity of the semiotic instauration in the face of its vulnerability and uncertainties. Moreover, the interpretative care intended to preserve the precarious complexity of the art object is merely the resumption of a primary task of symbolization: its retroactive significance, which has nothing to do with reproduction, with the representation of an already given order.

If we want to see art as a sensitive version (visible, tangible, sound, etc.) of the profound ideas of a culture, this aesthesic translation would only weaken their institutionalization. Art would only be an institution that proceeds in a direction contrary to institutional interests. As such, it also proceeds against itself, against its self-fulfilling institutionalization. Trying to escape its own "prisons" is a way of showing that there is further meaning beyond the law[14].

Nevertheless, art is not a carnival, a simple reversal of morals, but it reloads a *hiatus* that needs to be analyzed in depth: a *decoincidence* (Jullien 2017) that activates a critical separation between interpretative posture and the instituted horizon of living in order to bring about new conditions of *thinkability* (Ingold 2013, p. 304) and of sensitive apprehension.

8.3. Some remedies to institutionalized nominalism of art

Descola (2013a) believes that "Revisiting the question of the institution and stabilization of collective forms of experience thus becomes a matter of urgency" (p. 74). Today, after having classified and hastily dismissed structuralist experience, the awakening of "theory" in the human sciences seems to be going through the recognition of the heterogeneity of the ways in which values are established *and* the plurality of relationships between the positions of fabrication, management and evaluation. However, when Descola (2013a) speaks of the "the variety in forms of *worlding*" (p. 79), we may have the feeling of an indirect revival of Goodman's (1978) "ways of worldmaking". Naturally, it would be necessary to clarify how to take into account the different ways of constructing the world. But in any case, a review of the modalities of establishment cannot immediately translate into a description of the practices of value transformation and the characterization of *forms of life*. A typology based on establishment procedures can be fallacious because it also pre-establishes the criteria for classifying their results, without considering that

14 On the other hand, the "philosophical disenfranchisement" of contemporary Western art (Danto 1986) shows its paradoxical character: the theoretical claim to an autonomy of institutional functioning has led art to recognize itself in very thin conventions of existence and to encompass, formally and with the greatest indifference, all language games as if to celebrate their equipollence. The art supermarket then becomes a kind of pyre of semiotic vanities. As for the statements concerning the death of art, they are the most direct evidence of the fact that we are talking about an art that is no longer questioned by the aesthetic production of other cultures.

cultural difference must be measured in terms of practices, where systematic forms and mutually inconceivable interdomain relations are found before attempts at translation.

In other words, relativizing does not mean renouncing the use of one's own cultural dimension as a possible example of the human being, which must be integrated into other forms of expression. For example, in the tradition of comparative law, one manages to categorize equivalent responses to social problems because human forms of life cannot claim exclusivity of their condition. In humanity as an integrative composition of commensurable convergences and discords, the alternative cannot be "radical strangeness".

Can we generalize the considerations that we are developing around art from this negation of a radical strangeness? Does art have a structural vocation within a semiotic ecology? Beyond any idealization, does art signal a generalizable problem, a common paradox that each culture has tried to solve in a specific way?

However, comparative studies may try to answer these questions, but the premises discussed previously seem to be important in the search for promising results. For the time being, we would like to take a hermeneutic critique of art in contemporary society further, accepting neither its nominalist reduction and its framing within institutional procedures nor the rebirth of an idealistic, aestheticist or messianic vision. Here, then are some working hypotheses, with a prior restriction of the horizon of control (contemporary Western art), to then evaluate possible extensions of relevance.

8.3.1. *Art as a displayed vulnerability of institutions: maestria in minor mode*

What may have emerged as a form of general manifestation of the emancipatory vocation of culture, art, would question the dialectic between institutions of meaning and epicenters of creativity, working on gaps that promise further significations before the conquest of a full epistemic mastery of the field. Thus, the technical mastery of art (*synthesis*) is appreciated as inversely proportional to its possibility of self-observation of its inscription in a previously mapped terrain (*analysis*). Its experimental vocation is the result of a shift in the coupling between the instances promoting meaning and the reference *semiosphere* of a controlled skid leading to new balances and "prehensions". Always wanting to test new nodes of meaning transversal to other social domains, art does not dissociate aesthetic valences to free them from their connections with other "primary documents", but simply proposes alternative couplings or the recovery of old couplings, after the provocation an episode of cultural vertigo. The "re-coupling" is one with the symbolic effectiveness of art, its positional, cathartic and sometimes therapeutic power, beyond institutional oppositions.

Art can be seen as an exemplification of a symbolic equilibrium that displays the internal *play* of language games, beyond the performative social scopes, the programmatic discourses that characterize the spirit of the time. If art is a "deep play", its mission would be to test the very seriousness of games.

In contemporary Western societies, art seems to occur, on the one hand, under a concessive regime of divergence, differentiation of objectives and testimonies of alternative and marginal forms of life; on the other hand, it seems to be a strongly integrated field to protect general aspirations and in particular the ambition not to make us prisoners of our own games.

Behind facade idealisms, art has very limited powers, being reduced to the *maestria* of a resemiotization capable of drawing new couplings and new solidarizations between valences governed by different institutions. Beyond strictly aesthetic questions, in terms of a symbolic ecology, art would show the cultural assumption of a double bind: to value internal *difference* beyond programmatic functions and to suggest an increase in the rate of institutionalization in order to defend this internal production of difference. It is a way of saying that art invites us to institutionalize what can compensate for the neurosis of the instituted organization. But this can only have a paradoxical result.

Art would play against the simple administration of values. In this sense, faced with the distribution of the modal loads of programming discourses, art would display "reloading" paths that can find their foundation in the languages themselves. Resemiotization takes advantage of the constitutive vulnerability of languages, in the face of a perception that is insubordinate to any programmatic election of a plane of expression; as much as it takes advantage of the internal virtualities of semiotic systems, which are not only largely unexplored, but are also indicative of an organizational project that is still unfinished.

8.3.2. *Art as a fracture of proximity*

Our working hypotheses here have the task of criticizing initial assumptions: while today's emphasis is on modes of instauration, this theoretical emphasis risks overlooking the blanks in meaning and the institutional flaws that art constantly points out. As the promoter of an instituting power, art is also its critic, even its palinody. Art would then have the symbolic power to signal the costs or losses hidden behind institutions by displaying itself in a semiotic state even more vulnerable than ordinary practices of managing meaning. Art would signal endangered potentialities by embodying them. This means that the work of art would not be promoted as a cultural product that takes precedence over other semiotic productions, but as an example of what needs extra care, cooperation and acceptance in order to develop its meaning. Art as a "primary document" must be conceived above all as the risk of a "new beginning", and not as a claim to primacy.

At the bottom, in a celebrated artwork, there is also a reflection of a practice in constant suffering, with its failed attempts and forgotten products. Strongly anti-economic, the artistic project calls for a semiotic ecology based on heteronomous criteria in relation to the host organization, in order to promote its response and to envisage, despite the initial "irritation", "significant places" and previously unknown relationships with other "primary documents".

Using Rastier's (2001) theory of anthropic spaces, one might note that art proposes a *proximal*, almost intimate space, while having its foundations of legitimacy in a *distal* horizon. It turns its back on what has given it a social existence. But not only: art would not accept the vocation of societies to elect their ideals and represent them through idols. On the contrary, art *haunts* institutions with a divergent otherness that activates contiguous unexplored selections and under an unreconciled regime, i.e. it does not accept the same staging of coded practices. Art is a fracture of proximity that remains heterogeneous according to an *allopathic* regime, that is to say, the exemplification of a sensitive "other" constitution; an otherness non-negotiable because it is unavailable to be substituted by the spectator's body or to become a collective experience. Art digs into the antinomy of *sharing*: dividing and having in common, distributing roles and supporting a person in his or her feelings.

The act of weighing involvements and the aesthetic distance of art need to be institutionalized because the challenge is not to eliminate a fictitious, fractured proximity, a reflection dissociated from the institutionally displayed signs. A hyperesthesia, an unthinking activation of the senses, a homeopathic promotion could only reconstruct a direct implication and a ritual continuity, even agonistic, in relation to the social scene (in this sense, these experiences are outside the boundaries of art).

8.3.3. *Allopathic regime and the vulnerability of art*

Defending an autonomous domain of art and giving it edges, though porous, means distinguishing it from a diffuse aesthetic. Thus, art would propose itself as the intensive concentration of an aesthetic involvement that must make the gap between, on the one hand, the institutionalized meaning and, on the other hand, a call for an "other" meaning, i.e. sensitive. The latter belongs to us in a negative format of critical unavailability and open existential contiguity.

Certainly, the hetero-referential scope of art does not fail to articulate with the self-referentiality of art, its stratification of praxis, genres, forms of internal mediation, in short, an institutionalization of specific meaning. On the one hand, the problem of a homogeneous scenery, of an agonistic attitude between self-referential forms must be averted. On the other hand, the vulnerability of art in relation to other institutions must continue to be exemplified through respect for the interpreter's autonomy, for his "proximal" but distinct and singular enjoyment.

It is then necessary to insist on the opposition of art to communicative empathy; already in Aristotle's work, the *allopathic* regime of art is affirmed as a condition of true *catharsis,* which is not promoted by immediate adherence, but by the resistance of a mediating otherness. In art, this seems to be clarified through the refusal of the response of a "naked" flesh of a body (*Körper*) that does not yet have the cover of a proactive and fully assumed experience (*Leib*). This resistance to the aesthetic/empathetic fusion is motivated by the demand to obtain the critical encounter between forms of life and to show them, by way of example, the potential but also the vulnerability of their semiotic resources.

The Hamletic problem of art is the upstream choice of presenting a fictitious space to bear witness, in return, to an experiential truth or to indicate a repressed implication. And this testimony does not accept the limits imposed by other social domains.

In turn, the instituted institutions find themselves criticized by the presence of an institution – the art – that fails to go beyond its instituting desire. Indeed, in artistic production, there is always at least a partial discrepancy between the inherited heritage and the promoted forms; and the arbitrariness of languages is exploited as a terrain for a search for an "other", externalized motivation, oriented toward heteronomous care.

The state of health of the art world is paradoxically characterized by its need to continue to call for help, beyond complaints about its condition of institutional dependence. And yet, art, in its destitution, shows the institutionalizing spirit, its constitution of alternative worlds, although not immediately available for consumption. With the insidious proposal of a fractured contiguity, art obviously escapes from the evidence of the negotiation of meaning. As an institution of meaning resistant to appropriation, art possesses, in its institutional housing, the museum, a space of implementation that has the sole function of continuing to preserve its vulnerable unavailability.

Between abusive exploitation and oblivion, the artwork is penetrated by unresolved tensions that alone allow the preservation of its history and the topicality of its appeal. The ritual of the museum visit should in fact celebrate the anti-procedurality of the approach to the work, the circumstance of a specific relationship that withdraws into a space that is neither that of the institution nor that of the dialogue of private negotiation. The unavailability of the work has been interpreted as an *autotely*, a "monadological" concentration on one's own existence, or as the reflection of the artist's narcissistic aura. Against this romantic idea of an impenetrable depth of identity, the work on the other hand promotes a subversion of the spaces of circulation of values, a disassociation which, internal to traditional genealogies, may well symbolize the emancipatory capacities of a culture.

It is from this re-interpretation of contemporary Western art that we could now question a comparative aesthetics. Beyond ontologies, can we find in other semiotic ecologies practices that question the paradoxes of institutions of meaning by displaying the vulnerabilities and subsequent potentialities of their semiotic resources through "other" sensibilities? Are there products of these practices that function as *artworks*?

8.4. Some methodological remedies

8.4.1. *The work and its spaces of relevance*

Our working hypotheses have the objective of posing the question of art differently in a comparative framework. Through such hypotheses, the specific character of the current conception of art in the West can no longer be identified with institutional procedures or coded communicative functions.

But epistemological reflections must find a methodological horizon as well. This passage can be appreciated through the semiotic reconceptualization of the notion of *artwork*: indeed, the latter can appear to us as a text that negotiates the continuity of the modality of public implementation of the object that manifests its identity. Continuity seeks to guarantee respect for an interpretative approach to the detriment of free use. This means that the artwork stabilizes a public transmission of its identity through a pre-organized articulation between the discursive space of the text, the space of implementation that accompanies its modality of social presence and the space-time that has accompanied its material and statutory establishment.

Seeing the work of art in the relations between these three spaces of relevance (discourse, implementation, instantiation) is not an unnecessary complexification, but the attestation of the care it requires in the face of its constitutive vulnerability. Indeed, each space poses its resistances to the integrations suggested by the others, which explains why the identity of the work is not just a package of values to be deciphered, but rather a series of tensions between (i) linguistic rootedness, (ii) form of public circulation, and (iii) historical genealogy. This outlines the "form of life" of the work in which we can recognize identity polarizations that transform and resist over time through fields of integration: material, communicational and genealogical.

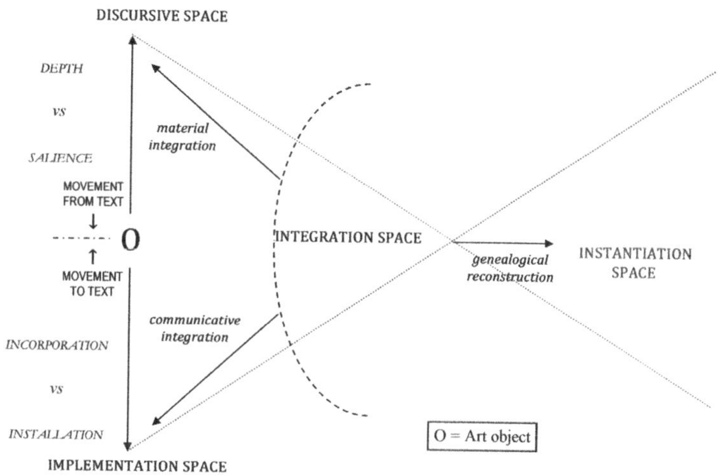

Figure 8.1. *Interpretation spaces*

In particular, it can be emphasized that a dialectic is established between the fields of integration so that the strong and restrictive codification of two identity polarizations is opposed, because of a system of compensations, to the density of another polarization. For example, an autographic work with a single copy (the genealogical integration of a painting that links it to the history of its production), subjected to a rigid implementation (the communicational integration of the painting into a decorative and evangelical plan of a church), does not fail to oppose such restrictions with an exaltation of the density of its iconic (pictorial) features, its texture, the material mediums and the discursive gestures (brush strokes), everything that makes the work singular.

But what can be noticed for the restrictions imposed by integration fields is also valid for the excessive freedoms they claim, freedoms that will then be compensated by other polarizations. For example, a plural autographic work, such as a lithograph, tends to place restrictions on its relevant textual qualities (in relation to its medias and the accidents of its preservation as a specimen) and its plural implementation in the same space is deemed abusive.

The relationships between the spaces of integration would deserve a more detailed description, but what interests us here is to present them as a critical framework for accessing the "form of life" of the artwork.

Obviously, behind the spaces of discourse, implementation and instantiation, one could recognize intentionalities; however, it is the system of multiple integrations that prevents intentionality from becoming a projection of unilateral and decisive meaning. The stakes of meaning are distributed and the "form of life" of the artwork is connected with different meshes of an open and evolving cultural environment.

8.4.2. *Cultural identity between analysis and interpretation*

Finally, we can propose a new distinction between *analysis* and *interpretation*: *analysis* can only separate investigation from discourse (semiotic enactment), implementation and instantiation, which implies the use of different methodologies and the crossing of disciplinary boundaries[15]. *Interpretation,* on the other hand, can only triangulate the fields of integration, through possible "anamorphosis" of its view of the work, which can be assumed as a *document* (dominance of instantiation), as a *discourse* (dominance of semiotic enactment) and as an *institutional object* (dominance of implementation). Anamorphic tensions and the three axes of negotiation never exempt interpretation from maintaining its integrative character in order to:

– look at the local organization of discourse through the general framework of implementation;

– take into account the past of production from an *archaeological* perspective (Foucault 1972) that starts from the "present" of the discursive traces of the semiotic enactement;

15 From the theoretical framework outlined above, it is clear that the open process of interpretation seeks to be "urbanized" through the prophylaxis of *analysis*. The latter can only begin if one has already "delimited" one's space of exercise (immanence), clarified the conditions of one's third-party gaze (explicit methodology) and guaranteed one's critical, even ethical (deontology), professional services. Analysis is only an expertise that momentarily blocks cultural transmission in order to make the semiotic nature of the transmitted goods locally appreciated; it guarantees a determination of meaning, which is within the expectations, but at the same time it condemns it to have no recoveries and no continuations. The analysis does not bear witness to a coupling with the inherited forms; on the other hand, it is obsessed with the *life of forms*, which it can approach only *a posteriori*, through *archaeological* studies (Foucault 1972).

– grasp the discursive formations that set up a space for implementation through the transmission of a cultural object rooted in a genetic past.

Naturally, we must admit that there are many tensions at play, but these remarks already lead us to change the traditional conception of "cultural identity"; the latter would be immanent to a single manifestation only through a combination of neutralizing effects of its possible reproduction and transcendence of manifestation. What always underlies such neutralizing operations is a coupling relationship that reveals the impossible isolation of any immanence. To illustrate this conclusion on the immanence of coupling, we can think of the question of fidelity to the unique copy of the work of art; although there is the concrete, or theoretical, possibility of a substitutive manifestation totally indiscernible from the original, the exclusive legitimacy of the authentic copy depends on the coupling with the author during the productive history of the work. In the same way, the possibility of accepting the fragmentary condition of a work depends on coupling with an uninterrupted chain of witnesses; the delimitation of the relevant copies for a multiobject sculptural work, or the number of legitimate variants of the same poetry, is acceptable only in coupling with the intentionality of the artist and the attestation of his/her performative acts.

The *autotelism* of the work of art, capable of forever setting the conditions of its experience, has been an aesthetic mirage that has sometimes become the exemplary model of the cultural object. On the other hand, the very defense of a material immanence that does not allow reproduction or reintegration depends on a series of coupling and neutralization operations.

8.4.3. *Methodology and knowing anew*

We can then end our development with the idea that a science of *art* is skeptical about the critical scope it should exemplify. Without doubt, science should not celebrate the work of art, but, if it does not have the vocation of maximizing the presumed values of the artistic object, it must recognize the dignity of the latter as an operator in history and in the present, which means that we must both philologically reconstruct the conditions of its form of life and understand the creative impulse to which it bears witness.

A science of art is called upon to be critical because its object of study is the exemplification of the paradoxical cohabitation of antinomic characters in cultures. The task of art is not to resolve paradoxes, but to attest how they are displaced and rearranged in cultures and in their historical phases. Only a science of complexity can deal with art, a science that seeks to reconnect links, unnoticed implications, not to deploy meaning, but to put it in tension. Its critical perspective is constructed in contrast to the transmission understood as a simple conveyance, a transport of information. This vision can ward off the temptation of the human sciences to propose themselves as an *ars magna*, a pragmatics of optimal conduct, with its analytical protocols as a unilateral, even exclusive, example of validated rationality.

On the other hand, if science can analyze works of art, art may well question the meaning of science. Aristotle stressed that "poetry, therefore, is a more philosophical and a higher thing than history" (Aristotle, *Poetics,* IX) as a science of facts. The *mimétikè poïésis* would have an interpretative primacy which, in our opinion, is in the paradox of a fiction that offers itself as a competitive, alternative model of coded self-understanding protected by an autonomous domain of regulated values. Obviously, art does not want to go beyond history, but only to anticipate a characterizing vision where each passage through the same cultural *topos* requires a reconfiguration, in short a new critical interpretation. Art reminds history of a critical requirement in the face of the coming present and the past, which must continue to be "born" in its most proper origin: and this does not happen through a search for purity, but through the recognition of instruments of objectification, mediation and internal modeling. By representation, imitation does not concern things, but the practice of knowing and producing them. One does not imitate the vase but the potter's art (Melandri 2014, pp. 55–56).

Melandri stressed that Vico is not Dilthey, and the distinction between natural and civilizational sciences is not the same as the distinction between *Natur and Geisteswissenschaften*. For Vico, the human sciences are not based primarily on epistemic principles, but on techniques and arts: this is the *scienza nova*. In the Yale seminars, Cassirer asserted that "we can regard the introduction to Vico's work as a new *Discourse on the Method*" (Cassirer 1988, p. 121, author's translation) and "there is no other field, he says, in which the human mind is closer to itself than in history" (Cassirer 1988, p. 121, author's translation).

The *factum* is not a *datum*, but what has been produced and experienced in a solidarity between *facticity* and *comprehension* (Melandri 2014, p. 58). Thus, science in the *humanitas* is only responsibility squared, *res gestae* that *insist* on other *res gestae*: double implications, increased focus on the *acted* and on the preconceptions that reveal their successes or their disasters. There is no renunciation of an apprehension capable of objectivizing relationships, but it must be considered that these relationships develop additional links.

Naturally, there are only "recoveries" of a truth to be understood, *mimesis* being necessarily a *dynamis*, but this recovery requires a *re-involvement* according to the two focal points of knowledge: scientific distance and the art of managing relationships, given the *interpenetration* between the observation system and the environment, between identity and otherness.

The term "resipiscence", suggested by Melandri, is interesting in this respect; it indicates a "*re*" (again) "*sipere*" (know), knowing anew, return to a correct view according to a progressive emendation of the text (singular) and a possible amendment of the laws (general). If we conceive art as a symbolic form, it should be stressed that this emendation is quite paradoxical, because it invites the recognition of a *methexis*, therefore of a participation, but only on the condition of having previously given an exemplification, suggesting an appropriation of values through an "other" *form of life*.

If a work of art is intended to have a passional efficacy, it must be based on an allopathic signification, i.e. a manifestation of affection mediated by a fictional alterity (the character is the "pathetic focus", not the actor as such on stage). In art as a symbolic form, separation and *stimmung* (common tonality) can cohabit as tensive and dualistic *semiosis*, framed by genres and registers that express differently the *cohabitation* of the singular and the institutional in a discourse with a constitutive polyphonic vocation. Art is always "bilingual", as Lotman liked to point out. If the result of art is not a sterile scepticism, in parallel, we can affirm that it will not be of the order of sophism, because the smallest possible world housed within the discursive framework, insubordinate to *linguistic government*, will be able to make us discover a core of meaning "offscreen" and thus affirm itself as a revealing *interpretant* of the discursive attitude itself.

8.5. References

Basso Fossali, P. (2002). *Il dominio dell'arte*. Meltemi, Rome.

Basso Fossali, P. (2011). Actualités esthétiques, questions sémiotiques. Quelques controverses autour du domaine de l'art. *Signata – Annales des sémiotiques*, 2, 81–120.

Basso Fossali, P. (2017). *Vers une écologie sémiotique de la culture*. Lambert Lucas, Limoges.

Basso Fossali, P. (2018). La sémiotique visuelle de Greimas entre archéologie et actualité. *La Part de l'Œil*, 32, 308–329.

Baxandall, M. (1972). *Painting and Experience in Fifteenth Century Italy*. Clarendon Press, London.

Boas, F. (1955). *Primitive Arts*. Peter Smith Pub Inc., New York.

Cassirer, E. (1944). *An Essay on Man*. Yale University Press, New Haven.

Cassirer, E. (1988). *L'idée de l'histoire. Les inédits de Yale et autres écrits d'exil*. Le Cerf, Paris.

Cassirer, E. (2004). *Aufsätze und kleine Schriften (1927-1931), Gesammelte Werke*. Hamburger Ausgabe, vol. 17. Meiner, Hamburg.

Cassirer, E. (2012). Form and technology. In *Ernst Cassirer on Form and Technology*, Hoel, A.S., Folkvord, I. (eds). Palgrave Macmillan, London, 15–53.

Cavell, S. (1987). *Disowning Knowledge: In Six Plays of Shakespeare*. Cambridge University Press, Cambridge.

Collingwood, R. (1938). *The Principles of Art*. Clarendon Press, Oxford.

Danto, A. (1993). *Philosophical Disenfranchisement of Art*. Columbia University Press, New York.

Descola, P. (2010). *La fabrique des images. Visions du monde et formes de la représentation*. Exhibition catalogue of 17th February to 11th July 2011, Musée du quai Branly – Somogy.

Descola, P. (2013a). *The Ecology of Others*. Prickly, Chicago.

Descola, P. (2013b). *Beyond Nature and Culture*. Chicago University Press, Chicago.

Eco, U. (1968). *La struttura assente*. Milano, Bompiani.

Eco, U. (1989). *The Open Work*. Harvard University Press, Cambridge, MA.

Elkins, J. (2017). Marques, traits, splendeurs. Pourquoi la peinture résiste à la sémiotique. In *Penser l'image III. Comment lire les images?*, Alloa, E. (ed.). Les presses du réel, Dijon, 75–116.

Foucault, M. (1972). *The Archaeology of Knowledge and the Discourse of Language*. Pantheon Books, New York.

Geertz, C. (1973). *The Interpretation of Cultures*. Basic Books, New York.

Geertz, C. (1983). *Local Knowledge. Further Essays in Interpretive Anthropology*. Basic Books, New York.

Geiger, M. (2005). *Zugänge zur Ästhetik*, 1928, *Vie all'estetica. Studi fenomenologici*. Clueb, Bologna.

Genette, G. (1997). *The Work of Art. Immanence and transcendence*. Cornell University Press, Ithaca/London.

Goldwater, R. (1973). Art history and anthropology: Some comparisons of methodology. In *Primitive Art and Society*, Forge, A. (ed.). Oxford University Press, London/New York, 1–10.

Goodman, N. (1968). *Languages of Art*. Bobbs-Merrill, Indianapolis/New York/Kansas City.

Goodman, N. (1978). *Ways of Worldmaking*. Hackett Publishing, Indianapolis.

Goodman, N. (1984). *Of Mind and other Matters*. Harvard University Press, Cambridge, MA.

Greimas, A.J. (1989). Figurative semiotics and the semiotics of the plastic arts. *New Literary History*, 20/3, 627–649.

Grosse, E. (1897). *The Beginnings of Art*. Appleton & Co., New York.

Grosse, E. (2009). *Les débuts de l'art*, 2nd ed. Esthétiques du Divers, Paris.

Ingold, T. (2013). *Marcher avec les dragons*. Zones Sensibles, Brussels.

Jullien, F. (2017). *Dé-coïncidence. D'où viennent l'art et l'existence?* Grasset, Paris.

Koselleck, R. (2011). *L'expérience de l'histoire*, 2nd ed. Gallimard/Le Seuil, Paris.

Lauschke, M. (2010). Les fonctions de l'art chez Ernst Cassirer. In *Ernst Cassirer et l'art comme forme symbolique*, van Vliet, M. (ed.). Presses universitaires de Rennes, Rennes, 27–41.

Levinson, J. (1979). Defining art historically. *British Journal of Aesthetics*, 19(3) 21–33.

Luhmann, N. (1992). Ökologie des Nichtwissens. In *Beobachtungen der Moderne*, Luhmann N. (ed.). Westdeutscher Verlag, Oplanden, 149–220.

Melandri, E. (2014). *I generi letterari e la loro origine*, 2nd ed. Quodlibet, Macerata.

Price, S. (1989). *Primitive Art in Civilized Places*. The University of Chicago Press, Chicago.

Rastier, F. (2001). L'action et le sens. Pour une sémiotique des cultures. *Journal des anthropologues*, 85–86, 183–219.

Rastier, F. (2016). *Créer: Image, Langage, Virtuel*. Casimiro, Paris.

de Saussure, F. (1996). Premier cours de linguistique générale. In *Les cahiers d'Albert Riedlinger*, Komatsu, E., Wolf, G. (eds). Pergamon, Oxford/Tokyo.

Wollheim, R. (ed.). (1980). Criticism as retrieval. In *Arts and Its Objects*. Cambridge University Press, Cambridge, 169–185.

Conclusion

Making Sense Between Science and Ethics

In a way, this conclusion is separate, less focused on strictly linguistic, semiotic or comparative epistemological problems encountered in the sciences. Indeed, it approaches the subject from the perspective of other types of knowledge, mainly those of moral and political philosophy. Ethical knowledge from moral philosophy and its derivatives in moral sociology, law or normative economics, as well as political theory, are untouched by the division between natural and engineering sciences and human and social sciences. It is undoubtedly separate because it falls under another epistemology of the sciences. Moral and political philosophy are about the "ought to be" rather than the "what is", or to put it another way, they are about the good rather than the true. This is what can give ethics an epilogue though it must be understood as "separate". Indeed, the epilogue is literally "above discourse". It has a special place.

However, this status can be questioned separately from ethics and political theory. For example, one may want to naturalize ethics with the experiments of moral psychology to try to understand how ordinary individuals evaluate this or that situation or dilemma.

Conversely, we can consider that ethics is not knowledge and that it is only irrational.

With these two types of positions, ethics, moral or political philosophy would have to answer some of the questions raised throughout this book.

Ethical questioning has an important part to play in the subject of meaning. Beyond understanding, making sense implies a normative dimension. Indeed, saying

---

Conclusion written by Bernard REBER.

that something or an action makes sense goes far beyond understanding and validating statements, but it is making a statement about the goodness or value that this thing or action entails. For example, it is nonsense to drive 4 × 4 cars in Paris, because they pollute and roads are tarmac everywhere.

This conclusion must be understood as both a discourse capable of recapitulating, or better still of reappraising part of the problems dealt with between making sense and making science, and an opening toward a continuation and a relaunching of the problem. In fact, many of the problems that arise, whether in science for political decision-making, in the case of health and the environment for example, public scientific controversies, for example energy choices or bioethical problems, or even what has been quickly classified as post-truth, require a decompartmentalization of the sciences of what is and of the sciences of what ought to be. This opening up is accompanied by the formidable problem of a crossed, interdisciplinary epistemology, but also on the two levels of truth and goodness or, to put it another way, of science and ethics. If these two requirements no longer exist, then it is difficult to make science and even harder to make sense.

This conclusion is therefore intended to address the question of how to consider together the dimensions that make science and those that make sense, understood as science and ethics. The gesture implies many reorganizations, epistemological choices and explanations that would have no place here for lack of space. We refer to the book *Precautionary Principle, Pluralism and Deliberation* (Reber 2016) for a complete reasoning. Here, we will only examine the conception put forward in political theory to bring together the demands of participation and rationality: deliberation, known as theories of deliberative democracy (TDD). Indeed, it is often on the side of politics, and more specifically of democracy, that one seeks the means to hold together all its dimensions. However, this theory often leaves out the problems of disagreements specific to political life, often based on interests, and proposes to justify the claims made during various kinds of debates. However, the notion of deliberation can also play a role at a meta-level: a deliberation between knowledge that can shed light on or document a situation, or even conflicts that are more traditionally political. We are at the crossroads between making science and making sense. It is not only interests or conceptions that are at stake, but also background knowledge.

In the Habermasian version of TDD, the most accomplished and in any case the most widely used, is a rhetorical notion, an argumentation, which is the crucial element in the process of mutual understanding and agreement to be found in situations of disagreement. Yet, while argumentation is a capital requirement for Habermasian deliberation, its role is even more important in his social science theory in order to save rationality from certain forms of relativism. We can therefore try to find out what conception he has of the argumentation he claims. The surprise is great if one consents to a careful reading of his major work, *The Theory of*

Communicative Action (Habermas 1984)[1]. One wonders whether the analysts who follow him in his requirement for argumentation have read it, for we shall see that its presentation is enigmatic, if not outlandish, to say the least. However, we will be careful not to lose the argumentative requirement in the deliberation, adding that it applies to particular situations. To put it another way, this communicative capacity has its own resources. In any case, for knowledge like philosophy and for its moral and political subfields, argumentation is one of the capital requirements. Surprisingly, however, here too it is not very well explained. We will therefore deal successively with deliberation, the theory of deliberative democracy, Habermas' argumentation, the subject of argumentation in an interdisciplinary context, and propose to conceive of it in the forms of *production*, *testing*, *contestation*, the search for *a common basis*, and even the search for *complements* to the missing parts of it. Argumentation more than other communicative skills is able to structure the inquiry.

C.1. The difficulties of inclusive deliberation

Social and political life often groups together science and meaning in a rough way. Political life is full of discourses that borrow from both registers, sometimes to describe, sometimes to evaluate. If one finds more or less understood and validated knowledge and more or less convincing justifications, the advantage of this environment is the obligation to hold together science and meaning. Even if the extremes of knowledge and decision-making are separate, institutionalized according to a social division of labor to which one owes great efficiency, political life, and more generally political communication occupies the main place and is at the crossroads between those who govern and those who are governed, and among the governed themselves. A movement to think broadly, to decompartmentalize and rebalance the various perspectives is necessary, certainly less wild and more controlled than discourse in public life. Often, it is the need for a broader inclusion of stakeholders (patients, affected audiences or even ordinary citizens) that is called for as a solution. This is the case for the experiences of Participatory Technology Assessment (Reber 2005 2011), or more recently of responsible innovation and research (Pellé and Reber 2016). Yet, inclusion is not a solution, as it brings with it other problems. All of these, together with the scientific, ethical and political dimensions, can be grouped under the following scope of questioning:

> How can we deliberate together, with a prior assessment from a large number of actors with different and asymmetrical skills and expertise (since we add the participation of ordinary citizens, experts and stakeholders), subject to rules of debate taken from democratic theories, when the stakes crystallize around innovative and

1 Shortened to TCA.

> controversial technologies, since they could cause serious and/or irreversible damage?

Beyond the problems arising from the constraints of the number of participants that can hinder deliberation or from the complexity of modern societies, this questioning can be reappraised in a more philosophical way, as follows:

> Deliberate with different ethical justifications (ranging from elements taken into account at the level of applied ethics, ethical theories and meta-ethical options), according to different political theories, but also with natural and engineering sciences and their disciplines, with the human and social sciences (HSS), with their fields of relevance (and therefore implicit exclusions), their modalities of producing evidence and dealing with uncertainties.

We can therefore see the two types of science appearing with this questioning, those concerning nature to move fast and the HSS. Similarly, each has partial and excluded knowledge. We could therefore hypothesize that they need to complement each other, which is what we shall try to defend in the last part, where the argument is conceived as an inquiry.

But before turning to the argument, let us look at how inclusive deliberation or the theory of deliberative democracy is imagined in the political realm, between political sociology and political theory.

C.2. Argumentation in the deliberation process

The theory of deliberative democracy, or more broadly, the important role of deliberation in politics, has become established in political theory (see, for example, Bohman and Rehg 1997; Chambers 2003). It is one of the most sophisticated and therefore most likely to accommodate a science and make sense. It should make it possible to resolve, or at least to manage in the best possible way, the conflicts arising from the plurality of interests, identities and conceptions that are expressed in our societies. It requires more than aggregated preferences as is the case in voting. It requires justifications for the claims made.

A first definition of this theory is given by Chambers:

> Deliberation is a debate and discussion aimed at producing reasonable, well-informed opinion in which participants are willing to revise their preferences in light of discussion, new information and claims made by fellow participants. (Chambers 2003, p. 309)

It should be noted here that Chambers returns to preferences while the TDD insists on the need for a reasoned discussion.

It is in this more Habermasian sense that another characterization of the TDD was proposed by the team of the political scientist Jürg Steiner, who outlined deliberative democracy according to a typology based on six criteria (Steiner *et al.* 2004)[2]. The summary is as follows:

1) arguments should be expressed in terms of "public good"[3]. If someone wants to assert their interests, they must be able to demonstrate their compatibility and contribution to the common good;

2) participants are required to express their views in a genuine and truthful manner;

3) participants are required to demonstrate a willingness to truly listen to the arguments of others and to treat them with respect;

4) the parties are required to put forward logical and valid claims and justifications, through an orderly exchange of information and good reasons[4];

5) participants are expected to be willing to follow the strength of the best argument. This means that their preferences should not be fixed definitively, but should be open to exchange and possible revision. This argument is not given *a priori*, but is to be sought during the joint[5] deliberation;

6) everyone participates at an equal level, without constraint, in an open political process.

This way of presenting the TDD is interesting because it gives prominence to arguments, which are the real central elements in the discussion, to achieve intercomprehension.

Previous work has already addressed the paradox that exists among both practitioners and analysts of deliberation, of requiring argumentation without defining it or simply giving it the characteristics to be able to recognize it. It is a deflationary position that involves the respect of participants that has been chosen since it is often this term that sums up the conception of practitioners and some analysts. This can be seen by using Steiner *et al.*'s six criteria. Argumentation is

2 Over time, Steiner changed his index, taking different paths (Steiner 2012).

3 John Rawls, claiming his contribution to TDD, in other words "public reason". For a discussion of the two Habermasian and Rawlsian versions of deliberation, see Reber (2016).

4 On this point, Habermas goes further with his belief in their universality.

5 Some authors go so far as to say that no one has any particular authority other than that of having a good argument to make.

more about the quality of the relationships between participants in the discussion than about a particular structure, content or form of communication, although the term argumentation appears in three requirements. Similarly, it has been argued that there is no need to choose between communicative skills, for example narration, interpretation or argumentation (Reber 2007). The empirical analysis of debates of the participatory technological assessment (PTA), gathering a diverse group of citizens and experts who seek to make science while making sense, shows more radically that below these communicative capacities we encounter the problem of the constitution of the question and the adequacy of the answer(s). This problem is not insignificant, since it is often found to be lacking in experiments of the PTA type. Similarly, the answers are difficult to grasp in a competition of pretensions, as is the case in parliamentary debates.

In this conclusion, on the contrary, it will be shown how argumentation is a communicational capacity of its own, even irreplaceable in certain circumstances, particularly in order to confront the pluralism of moral theories (Reber 2006). It should be added that the argumentation framework must be respected, at the risk of ridding the TDD of its meaning. This position goes against those who leave argumentation as optional and all those who do not bother to specify what is meant by argumentation. In the deliberative subdomain, as far as we know, no one has bothered to do so. Moreover, even Habermas, when asked for support and/or caution to be exercised, is not much more precise; worse, he loses us, as explained below.

C.3. The ghost of the Habermasian argument

To find out what Habermas had in mind when he spoke about argumentation, we have to go back to the TCA. As indicated, Habermas and Rawls are often used as sources and as authoritative arguments, and therefore the weakest argument, for those who speak of deliberation, whether practitioners or even political theorists. While it is doubtful that they have read or understood them, they never return to the conception of the argument defended by Habermas, nor do they try to develop one. Rawls' conception will not be discussed here[6], but it should be noted that in his famous debate with Habermas he was not convinced by the need to argue and did not understand what the latter meant by argumentation. It is Habermas in particular who inspired the work of Steiner *et al.* In some ways, they are very courageous, even reckless. Although, Habermas is anything but clear on this issue, which is central to his TCA opus. Indeed, the argumentation in this work is more broadly the condition of rationality. The theory of society he proposes is in dire need of a theory of argumentation to make the concept of communicative rationality explicit

6 Rawls's problem is above all that of reasonable pluralism, which he only approaches to a point, subjecting ethical pluralism or political pluralism (see Reber 2017).

(Habermas 1984, p. 18). He addresses the problem of argumentation in an excursus, whose structure alone would merit an analysis in terms of argumentation. For example, on several pages he uses an almost unknown author, Wolfgang Klein, a specialist in argumentation, to criticize Stephen Toulmin, whose definition of the components of argumentation is one of the most important in contemporary philosophy. He then uses Toulmin to criticize Klein, who is considered too contextualist in his search for the effectiveness of an argument. In addition, he often uses Toulmin's definitions on important points, such as the famous Toulminian definition of the unconstrained constraint of argumentation. It is the key element in settling conflicts, changing one's position and reaching forms of agreement, including consensus in Habermas.

He says of argumentation:

> It is that type of speech in which participants thematize contested validity claims and attempt to vindicate or criticize them through arguments. (Habermas 1984, p. 18)

We notice a problem, or a gray area, with this first definition, since argumentation is a type of discourse, when arguments are components of this type of discourse. Argumentation is therefore broader. However, the problem remains to know what these components are. Moreover, if argumentation is not the equivalent of an argument, before which force should we yield? It seems that it is arguments since argumentation is the discursive whole in which the participants are included. They will therefore have to be defined.

At the very least, it is awkward to put back into the definition of argumentation terms that are close to arguments. We will need a theory of argumentation as a type of communicative ability that must be distinguished from the components of arguments themselves.

Further on, Habermas thus completes the description of the argument along a very simplified, partially Toulminian line: "An argument contains reasons or grounds that are connected in a systematic way with the validity claim of a problematic expression."

The vagueness of these two outlines of a definition can be explained by Habermas' personal addition in what he calls "the outer court of the theory of argumentation", which is intended to supplement his "provisional definitions of the concept of rationality" (Habermas 1984, p. 43).

Indeed, Toulmin, Klein and a symposium on informal logic that took place in 1980 provide the main elements to this partial theorization of argumentation. We would only be in the vestibule. But then why not enter the whole house and visit argumentation in all its aspects, or at least the main rooms to complete his metaphor, all the more so as it is a question for him of saving rationality, which he qualifies as communicational and opposed to a strategic rationality?

In addition to this lack of clarity and overly introductory statement, Habermas complicates matters with this personal addition.

He proposes three aspects of argumentative discourse as (a) process, (b) procedure and (c) argument production.

a) This aspect can be summarized by the argumentation included:

> as a reflective continuation, with different means, of action oriented to reaching understanding. (Habermas 1984, p. 25)

He goes on to say:

> a form of communication that is improbable in that it sufficiently approximates ideal conditions. (Habermas 1984, p. 25)

This is bad news for those who do empirical work on real discourses since this kind of communication is unlikely. However, we will answer that it is possible to measure proximities with the ideal, to do one's best in institutional *design* to tend toward it, and that in philosophy it is not impossible to aim for ideal conditions, especially in normative matters. Habermas acknowledges this for himself:

> I tried to delineate the general pragmatic presuppositions of argumentation as specifications on an ideal speech situation. (Habermas 1984, p. 25)

Specifically, he understands the argumentation process to mean:

> reconstruct the general symmetry conditions that every competent speaker must presuppose […] to enter into argumentation. (Habermas 1984, p. 25)

In addition to the symmetry of the relationship, he adds the exclusion of any constraint except that of the better argument. The only valid motive is that of "a cooperative search for the truth" (Habermas 1984, p. 25).

b) He understands argumentation as procedure as a form:

> [...] with a specifically regulated "cooperative division of labor between proponents and the opponents."

He goes on to say:

> participants
>
> - thematize a problematic validity claim;
>
> - relieve of the pressure of action and experience, in a hypothetical attitude;
>
> - test with reasons, and only with reasons, whether the claim defended by proponents rightfully stands or not.

c) With respect to the production of arguments, he has in mind the relevance of those arguments "by virtue of intrinsic properties". They are the:

> means by which intersubjective recognition of a proponent's hypothetically raised validity claim can be brought about and opinion thereby transformed into knowledge. (Habermas 1985, p. 25)

He then recalls what he says are the components of the arguments as proposed by Toulmin, omitting the modalities. A comparison between Habermas' presentation of them (Habermas 1984, pp. 25–26) and the quotation he takes up (Habermas 1984, pp. 410–411) could be the subject of an article. It is not certain that Toulmin recognizes himself in it.

We will end this with a few remarks.

First, it is not understood why argumentation as a process (a) or as a procedure (b) should be distinguished. Moreover, authors who have done empirical work based on Habermas' theory have not used this distinction.

Second, to honor point (a), that argumentation is a "reflective continuation, with different means, of action oriented to reaching understanding" (Habermas 1984, p. 25), other communicative skills may well be candidates. Argumentation does not have a monopoly on reflexivity or on a specific reflexivity that should be developed.

Third, point (a) refers to argumentation, and point (b) refers to a claim and thus an argument.

Fourth: why these three analytical aspects (a, b and c) – of which there is doubt as to their relevance – given that Habermas himself believes that "the separation of the three analytical levels cannot be maintained"? (Habermas 1984, p. 26)

This tripartition becomes bizarre when he considers that it contains the canonical Aristotelian disciplines: rhetoric for (a), dialectics for (b) and logic for (c). It is tempting, but rhetoric itself includes a theory of argumentation alone. Argumentation differs from dialectics, which is based on demonstrations and not arguments. In these matters, it is better to follow Chaïm Perelman and his new rhetoric.

Since the problem of differences in intensity between arguments is crucial (Habermas 1981, p. 5), why did Habermas fail to include in the components of the argument the arguments conceived by Toulmin, that is of modalities? Indeed, they play an important role in the intensity of arguments. Yet, although this is present in Habermas' note 26 (Habermas 1984, p. 411), he did not consider it necessary to include this crucial element, whether in science, in political debates about the future, or in the interaction between science and political decisions, in the case of precaution for example (Reber 2016).

In the French introduction, 5 years after the German version, he accepts a weakness and acknowledges a limit to these two massive volumes. His first words are:

> I am aware that, despite the title, I do not submit with these two volumes a fully elaborated theory. (Habermas 1981, p. 9, author's translation)

However, this lack of elaboration relates in particular to the argumentation, which is central to its communicational action.

> For this reason I believe that the concept of communicative rationality, which refers to an unclarified systematic interconnection of universal validity claims, can adequately explicated only in terms of theory of argumentation. (Habermas 1984, p. 18)

Here he speaks about "The medium in which we can hypothetically test whether a norm of action, be it actually recognized or not, can be impartially justified is practical discourse; this is the form of argumentation in which claims to normative rightness are made thematic." (Habermas 1981, p. 35)

This theory, if it can be traced back to Aristotle, is "still in its beginnings" (Habermas 1984, p. 22). It is a matter of informal logic.

Unfortunately for us, and for those who rely on Habermas in thinking that one should argue in controversies or public debates, Habermas in his development did not go further on this issue. He would even have yielded to a form of contextualism

(Thomas-Fogiel 2005, p. 132s). We could therefore say that he succumbed to what he reproached Klein for.

Habermas' supposed argument for moving from opinion to knowledge falls short of its promise.

Should we give up then? We do not think so. An attempt to show how for certain discussions a dialogical argumentation in the Toulminian mode is possible has been made elsewhere (Reber 2017, Chapter 6).

C.4. Arguing in an interdisciplinary context

Let us start again with Toulmin to identify the structures of an argument. It is:

> like an organism. [...] It has both a gross, anatomical structure and a finer as-it-were physiological one[7]. (Toulmin 2003, p. 87)

It may occupy several printed pages and may require a "15-minute oral description".

This assertion alone either dulls the requirement to argue, invoked at the thresholds of public debates such as the *Commission (française) nationale du débat public* (CNDP) (French National Commission for Public Debate) for example, or indicates that the much too short time frames stifle the possibility of arguing. It would be possible to argue, but with the goals to really develop the arguments by arranging times dedicated to each element of a complete argument.

Despite the difficulties of the strength of disciplinary fields and the length of the development of an argument, Toulmin proposes to recognize "a good half-dozen functions to be performed by different sorts of proposition" (Toulmin 2003, p. 131). His outline includes the following elements: data (D) (which are explicit and general), warrants (W) (which are often implicit), grounds (F), modal qualifiers (Q), forgotten by Habermas, conditions of exception or rebuttal (R) and conclusions (C) (Toulmin 2003, p. 93s).

Our problem is that in the PTA experiments and most public debates involving controversial scientific issues, the juxtaposition of disciplines is no longer tenable. Indeed, the problems to be solved are rarely within the understanding of a single body of knowledge. The Toulminian conception of argumentation therefore poses a problem with regard to interdisciplinarity. Certainly his perspective was motivated by resistance to the reduction to an analytical model of argumentation, especially

7 The level of logical forms.

against Ross. Indeed, he wrote to conclude his famous chapter on the organization of arguments:

> Many of the current problems in the logical tradition spring from adopting the analytic paradigm-argument as a standard by comparison with which all other arguments can be criticised. [...] Analytical arguments are a special case, and we are laying up trouble for ourselves, both in logic and in epistemology, if we treat them as anything else. (Toulmin 2003, p. 134)

On the one hand, against Habermas' universalism, Toulmin's position is conscious of the relevance of arguments according to their disciplinary fields, but, on the other hand, he confronts us with the difficulty of their compartmentalization, when the problems to be dealt with concern several disciplines. We therefore find a partial form of demarcationism in it. However, this position poses a problem since the coexistence of arguments is not always possible, since the arguments partially draw their strength from their disciplinary contexts, what Toulmin calls fields. Everything is therefore, in part, to be developed according to the case, since it is from there that the arguments draw their strength. In this way, the conception of the argument as an investigative device will be proposed.

Moreover, the sciences can complement each other, but also sometimes conflict (Reber 2017). In the first case, it is for example the complement between climate science and economics to know what the impacts are between incentive measures, for example the carbon tax, and atmospheric pollution. In the second case, it is the opposition that there may be between this type of incentive and certain ethical conceptions that are more sensitive to equity values or to the intrinsic value of air quality and its limits in terms of resilience.

We will therefore present, more briefly than in *Precautionary Principle, Pluralism, Deliberation. Science and Ethics* (Reber 2016), two attempts to address the subject of argumentation in an interdisciplinary context. Jacques (1989) and Quinche (2005) proceed by resorting to dialogism, believing that the Toulminian outline is too solipsist (somebody thinking alone). Indeed, according to them, the argument can be constructed in the interindividual interaction in the course of questions and answers. The first one transforms dialogically the Toulminian argumentative model by introducing a question and an answer at each stage of the argument. It stakes out its pattern of questioning that an interlocutor, virtual or not, could put forward. We are therefore witnessing a real questioning in an almost material sense of the questioning task.

With Toulmin, Jacques accepts that formal validity in the sense of classical logic is not a sufficient condition since the validity of an argument depends on the nature of the problems. To go further, he proposes to distinguish between the modes of interrogation[8] of science, anthropology, art, religion and philosophy.

Now, with ethical argumentation, in the case of ethics applied in bioethics discussions (Quinche 2005, pp. 148–157) or in places where it is a question of making science and making sense at the same time, it is necessary to decompartmentalize the disciplinary territories in order to respond to the investigation carried out by the participants within these bodies, which are in fact interdisciplinary, both in their composition and in their investigations. She argues that:

> The ethical argument would be [...] even more dialogical than presented by F. Jacques [...] because it would have to expose the plurality of sets of norms as well as the possible worlds that could be envisaged. (Quinche 2005, p. 155, author's translation)

Quinche then proposes the different stages of a questioning that she qualifies as ethical (Quinche 2005, pp. 46, 102, 405s): (1) common questioning of the situation; (2) reconstruction of the problem; (3) exploration of the levels of the possible; (4) normative level; (5) deontic level[9]; and (6) practical applications.

We emphasize that this promising conception does not challenge the science. In our view, it may be overly optimistic in its ethical perspective:

> A process where ideally the parties concerned jointly determine the issues and solutions. Ethical language is then no longer the instrument of resentment and accusation, but already constitutes the first steps towards reconciliation or even a common act of reformation of the referent. (Quinche 2005, p. 102, author's translation)

We do not share to the end this monopoly attributed to the ethics of arbitration and the organization of the debate of disciplines. Moreover, this monopoly makes the strong hypothesis of a possible collective reconstruction of data. It is preferable to give ethics its rightful place, but that its place, in interaction with other knowledge, should be without absolute priority. A type of interaction between ethical and "factual" aspects should be considered according to a codependence. Indeed, it is in a reciprocal questioning between ethics and science that both can advance. Let us refer to Reber (2017), but to give an example of the new technological possibilities that can advance ethical reflection or call upon it, for

8 Strictly speaking, argumentation is not a mode of questioning but rather a proposal with its justifications.

9 It thus calls for obligations and the success of the best actions.

example with the adage that it is medicine that has saved ethics by addressing new problems to it, at the origin of the prodigious development of bioethics. This trend has now shifted to the ethical problems of the digital age.

Conversely, ethics, if it is not to be a science fiction ethics, must review its forms of reasoning in the light of scientific developments in order to reason on proven bases and not on fantasies or promotional discourses that are far removed from reality. Ethics not only *exposes* but *is exposed to* the plurality of sets of standards or procedures from other disciplines, especially those that provide and validate results.

C.5. Argument as an investigation

So far we have seen with Habermas the recognition of symmetry, that is his term, in communicative interactions. He argued this long before the constructivist sociologists of science, who used the term at every turn. Habermas adds to the exclusion of any constraint except that of the better argument, since the only valid motive is that of the cooperative search for truth. However, we have seen that he leaves the constitution of arguments in the shadows. The proposals in the previous section are more precise.

Nevertheless, the arguments should be explored by combining internal and external characteristics of the arguments. To this end, five ways of conceiving arguments are proposed, integrating the Toulminian components while moving away from the irenic versions of previous dialogisms. Indeed, arguments can be considered differently depending on whether one *produces* them, *tests* them, *challenges* them, wants to *share a common basis* or *completes* them. This will constitute a document to be added to the record of deliberative democracy by going beyond the mere respect of arguments but to their discussion or even their proper rejection.

1) *Producing*: in the most minimal form (a), it is a matter of properly stating the argument that one has prepared to justify one's position. It may in some cases prove to be the most complete form if it is ultimately found that the argument prevails.

2) *Testing*: however, the argument sometimes needs to be tested. It is not only greeted with respect, it is also subject to scrutiny (b).

3) *Contesting*: this can go as far as to contradict. This can be exercised on each of the elements, the inferences (c), or in the comparison with (c') to another argument (that of another protagonist).

This stage with its two forms is already a first dialogic conception. The situation (c') is likely to be frequent since one of the buzzwords of the theory of deliberative democracy is to put forward one's arguments.

The difference between the two arguments may vary, not only on inferences or different choices in the structure of the argument but also because of the premises and data. The latter are often different[10] without explanation. They are a great source of disagreement that could prove to be overwhelming, in a deceptive way, since we are not really talking about the same thing.

4) *Sharing a common basis*: one can therefore enter into a deliberation on precisely this explanation in order to (d) defend a starting point (premises and data). We try to share a first common ground.

5) *Completing*: it may happen that we are not simply dealing with a work of producing, testing, contesting or clarifying the starting points, or even other lines of reasoning being argued[11], but that complements participating in the construction of the argument are requested.

There are two possible scenarios: either other participants have the information that is missing from our argument (e), or we need to go or push the survey further to get the missing information (e'). This means, for example, that his argument is too uncertain on some of its links, or undecidable in the event of opposition between protagonists on one of the steps compared in the two arguments.

A third party will supplement or even impose by the reply a reconsideration of one or other argument or their components.

We are therefore dealing with a deepening or solidification of the argument. We also point out that the argument is a tremendous incentive to investigate. It can also participate with its different forms in a decontextualization that other communicative capacities are more radically successful.

The last argumentative form can invite other more or less suitable skills to enter into its structuring. We must therefore go beyond the simple legitimacy of the participation of individuals through their competences as per the investigation.

It will thus have been understood that reality is rarely encircled by a single discipline, but rather by the summoning of a plurality of expertise that is required to

10 In the data, the premises or their relationships.

11 In order not to complicate the presentation too much, each of the possible bifurcations corresponding to the different elements that one chooses to recognize in a complete argument are not developed. Likewise, no details are given of the levels of certainty of each of the stages of the argument, or even of the questions relating to the enthymemes, eluding one of the premises or steps of reasoning held for some. Argumentation theorists disagree among themselves. If we take the Toulminian structure, some add steps or questions like Jacques, others on the contrary eliminate them, like Adam (2004).

make sense by making science and vice versa. We are then moving toward an interdisciplinary co-argumentation.

C.6. In summary

Various communication skills are possible, such as different ways of "making sense". This notion refers to the pragmatics of Apel's communication to which Habermas acknowledges his debt, before parting with it. Among these capacities, argumentation is promoted, both in the theory of deliberative democracy and in Habermas' communicational action. However, there is frustration with Habermas when trying to clarify what he means by argumentation. Moreover, he muddies the waters by displaying skepticism about the work of Toulmin and Perelman, while using them, sometimes with surprising rewordings. As has been shown, the clarity of his definitions, which oscillate between argumentative discourse, argumentation and arguments, is inversely proportional to the authority he brings to bear on argumentation to save rationality in the processes of intercomprehension and agreement. Habermas' greatest effort to indicate what he means by argumentation appears in a series of programmatic notes to found an ethics of discussion, in a debate between skepticism (epistemological and semantic) and cognitivism, where he sticks to the demand for the expected universality of argumentation. However, the outlines of argumentation need to be more precise in the TDD, both for political theorists and empirically oriented politicians who need to know what they are looking for. Those who have tried, like Steiner *et al.* (2004), have taken up Habermas' work. However, the excerpt from Habermas' more recent book *Between Facts and Norms. Contributions to a Discourse Theory of Law and Democracy* (1996), taken as a guarantee by Steiner, already indicates a problem. Communicative action refers to an "argumentation process" involving the need to "justify their validity claims before an ideally expanded audience" (Steiner 2004, p. 21; Habermas 1996, pp. 322–323). The specific nature of argumentation would only be to allow for going "beyond local practices of justification and to transcend the provinciality character of their spatiotemporal contexts" (Steiner 2004, p. 21; Habermas 1996, pp. 322–323).

On the one hand, an intercultural semantics is possible without going through argumentation (Pharo 2004). On the other hand, above all, this conception remains vague.

Steiner *et al.* therefore had to look elsewhere for a semiotic definition of argumentation borrowed from Sebeok to constitute their *Index de la qualité du discours* (Discourse quality index), particularly under the heading of *discourse* justification. We note in passing that they talk about discourse and not deliberation and that Steiner (2012) has rearranged this point in the book, *The Foundations of*

Deliberative Democracy. This book echoes a wealth of empirical work on TDD that tries to take into account another communicative capacity–storytelling. Storytelling is said to have certain virtues that argumentation does not have, in particular the fact that it can better account for identity-related demands and is better able to deal with emotions or the restitution of testimonies. It is therefore necessary to recognize, alongside argumentation, a place for narration within deliberation.

Another important communication capacity that has been little explored within the TDD is interpretation. Interpretation is frequently used, particularly in the legal world, which has been used since Aristotle as a framework for argumentation and deliberation, notably in his *Rhetoric* (Reber 2017). It is the interpretation that Ronald Dworkin promotes as the best interpretation to be found in the discussion of cases (Reber 2013). His strategy is close to that of Habermas' strategy. It aims to "pretend" that this best interpretation exists. It is supposed to guarantee justice in law and equal consideration of individuals. A similarity with Kantian teleological judgments can be recognized. Moreover, in Habermas' case, the kind of consensus behind his reconstruction has often been misunderstood. It is not a real but a virtual agreement, or to use his words, he is in an ideal situation of interlocution. It should be added that after agreement, argumentation would no longer be necessary, hence Habermas' refusal to define argumentation in any way other than by taking over the rules of law from Robert Alexy, who had written his thesis on legal argumentation. However, in a trial one arrives at more clearly identifiable closures and decisions than in public debates. The Habermasian consensus is a bit of an "as if consensus was possible" consensus, which means that actors agree to participate and strive for it.

However, other serious candidates and defenders of interpretation, more open to the variety of interpretations, exist: Gadamer, Husserl and in a more Aristotelian line of singular judgment, Paul Ricoeur. Interpretation would thus be a communicative capacity to be put forward within the TDD.

If we accept that deliberation is not a communicative capacity for itself, but it consisted of the three capacities mentioned, it will first be necessary to distinguish between them and then to see what role they can play in deliberation. Due to lack of space, work on each of them will not be undertaken. To go quickly, in discursive mode, the *narrative* focuses on descriptions of facts and events (Ferry 1991). The *interpretative* refers to explanations, based on causes, to search for meaning. It can provide interpretative frameworks or models. *Argumentation* gives rise to justifications consisting of reasons (calling for validity). It can be based on syllogisms (Ferry 1991), but many other forms are recognized as argumentation (Van Eemeren *et al.* 1996; Quinche 2005).

With regard to the different communicative capacities, the chords, without choosing a conception of them here, will be different depending on whether we are talking about narration, interpretation or argumentation. This is also true for the radical starting point mentioned above, i.e. the constitution of the question to be agreed upon and the adequacy of the answer(s). It will be necessary to agree on the description of facts for the narrative, their causes and meanings for the interpretation, and their justifications recognized as valid for the argumentation. The different communication capacities skills presented in this way complement each other. Moreover, narration and interpretation could be integrated as two moments of the argumentation understood according to the Toulminian sense. Agreements would therefore be expected at each of the moments of the argumentation to take up our formulation.

The institutional design of public debates could support the argumentation, but one can very well imagine a breakdown between institutions of the argumentative moments according to the relevance of each of them.

We have tried to clarify the resources of argumentation in relation to other communicative skills, such as narration and interpretation. Argumentation, when it is complete, includes different moments, which can be explained, evaluated, and which may or may not lead to agreements. Argumentation that is not conceived as a weapon for contradictory debates, but to be constructed, would reduce asymmetry, where it is often the implicit and the reasons imposed as obvious, when they are false, that benefit the weakest audiences. In this way, fears of inequality in the communication capacities privileged in the TDD are addressed, while retaining the benefits of the virtues of argumentation, both epistemic and normative.

C.7. References

Adam, J.-M. (2004). Une approche textuelle de l'argumentation : "schéma", séquence et phrase périodique. In *L'argumentation aujourd'hui*, Doury M., Moirand, S. (eds). Presses Sorbonne Nouvelle, Paris, 77–102.

Bohman, W., Rehg, W. (eds) (1997). *Deliberative Democracy. Essays on Reason and Politics*. MIT Press, Cambridge.

Chambers, S. (2003). Deliberative democracy theory. *Annual Review of Political Science*, 6, 307–326.

Ferry, J.-M. (1991). *Les puissances de l'expérience II*. Le Cerf, Paris.

Habermas, J. (1984). *The Theory of Communicative Action. Reason and the Rationalization of Society I*. Beacon Press, Boston.

Habermas, J. (1986). *Morale et communication. Conscience morale et activité communicationnelle*. Le Cerf, Paris.

Habermas, J. (1996). *Between Facts and Norms. Contributions to a Discourse Theory of Law and Democracy*. Translated by William Regh. MIT Press, Cambridge, MA.

Jacques, F. (1989). Dialogisme et argumentation : le dialogue argumentatif. *Verbum*, XII(2), 221–237.

Pellé, S., Reber, B. (2016). *From Ethical Review to Responsible Research and Innovation*. ISTE Ltd, London, and Wiley, New York.

Pharo, P. (2004). *Morale et sociologie. Le sens et les valeurs entre nature et culture*. Gallimard, Paris.

Quinche, F. (2005). *La délibération éthique. Contribution du dialogisme et de la logique des questions*. Kimé, Paris.

Reber, B. (2005). Technologies et débat démocratique en Europe. *Revue Française de Science Politique*, 55, 811–833.

Reber, B. (2007). Entre participation et délibération, le débat public et ses analyses sont-ils hybrides du point de vue des théories politiques? *Klesis. Revue philosophique, Philosophie et sociologie*, 6/1, 46–78 [Online]. Available at: http://revueklesis.org/index.php?option=com_content&task=view&id=45&Itemid=63.

Reber, B. (2011). *La démocratie génétiquement modifiée. Sociologies éthiques de l'évaluation des technologies controversées*. Presses de l'université de Laval, Quebec.

Reber, B. (2016). *Precautionary Principle, Pluralism, Deliberation. Science and Ethics*, ISTE Ltd, London, and Wiley, New York.

Reber, B. (2017). Dworkin est-il un réaliste moral et un adversaire sérieux du pluralisme moral ? In *L'empire des valeurs*, Policar, A., Dworkin, R. (eds). Classiques Garnier, Paris, 229–252.

Reber, B., Seve, R. (eds) (2006). *Le pluralisme. Archives de philosophie du droit*. Dalloz, Paris.

Steiner, J. (2012). *The Foundations of Deliberative Democracy. Empirical Research and Normative Implications*. Cambridge University Press, Cambridge.

Steiner, J., Bächtiger, A, Spörndli, M., Steenbergen, M.R. (2004). *Deliberative Politics in Action. Analysing Parliamentary Discourse*. Cambridge University Press, Cambridge.

Thomas-Fogiel, I. (2005). *Référence et autoréférence. Étude sur le thème de la mort de la philosophie dans la pensée contemporaine*. Vrin, Paris.

Toulmin, S.E. (1993). *Les usages de l'argumentation*. PUF, Paris.

Van Eemeren, F.H., Grootendorst, R., Henkemans, F.S., Blair, A.A., Johnson, R.H., Krabbe, E.C.W., Plantin, C., Walton, D.N., Willard C.A., Woods, J., Zarefsky, D. (1996). *Fundamentals of Argumentation Theory. A Handbook of Historical Backgrounds and Contemporary Developments*. Lawrence Erlbaum Associates Publishers, Mahwah.

List of Authors

Pierluigi BASSO-FOSSALI
University of Lyon 2
France

Vincent BONTEMS
CEA
Paris-Saclay
France

Astrid GUILLAUME
Sorbonne University
Paris
France

Lia KURTS-WÖSTE
University of Bordeaux
Pessac
France

Jean LASSÈGUE
CNRS – (National Centre for Scientific Research)
Paris
France

Régis MISSIRE
Université Jean Jaurès
Toulouse
France

Franck NEVEU
Sorbonne University
Paris
France

François RASTIER
CNRS (National Centre for Scientific Research)
INALCO (National Institute of Oriental Languages and Civilization)
Paris
France

Bernard REBER
CNRS (National Centre for Scientific Research)
Sciences Po (Political science)
Paris
France

Arild UTAKER
University of Bergen
Norway

Index

Only the first instance of "culture" and "semiotics" in each chapter has been listed in this index.

F

G, H

I

K, L

M

N, O

P

S

T

V, W, Z

Other titles from

iSTE

in

Interdisciplinarity, Science and Humanities

2019

ANCORI Bernard
The Carousel of Time: Theory of Knowledge and Acceleration of Time

BOULEAU Gabrielle
Politicization of Ecological Issues: From Environmental Forms to Environmental Motives

GAYON Jean†, interviewed by PETIT Victor
Knowledge of Life Today: Conversations on Biology

KHATCHATOUROV Armen, with the collaboration of CHARDEL Pierre-Antoine, FEENBERG Andrew, PÉRIÈS Gabriel
Digital Identities in Tension: Between Autonomy and Control
(Innovation and Responsibility Set – Volume 5)

LENOIR Virgil Cristian
Ethically Structured Processes
(Innovation and Responsibility Set – Volume 4)

NIKOLOVA Blagovesta
The RRI Challenge: Responsibilization in a State of Tension with Market Regulation
(Innovation and Responsibility Set – Volume 3)

ZEMBRI-MARY Geneviève
Project Risks: Actions Around Uncertainty in Urban Planning and Infrastructure Development

2018

LOISY Catherine, COULET Jean-Claude
Competence and Program-based Approach in Training: Tools for Developing Responsible Activities
(Innovation and Responsibility Set – Volume 1)

PANSERA Mario, OWEN Richard
Innovation and Development: The Politics at the Bottom of the Pyramid
(Innovation and Responsibility Set – Volume 2)

2017

MAESSCHALCK Marc
Reflexive Governance for Research and Innovative Knowledge
(Responsible Research and Innovation Set – Volume 6)

PELLÉ Sophie
Business, Innovation and Responsibility
(Responsible Research and Innovation Set – Volume 7)

2016

GIANNI Robert
Responsibility and Freedom: The Ethical Realm of RRI
(Responsible Research and Innovation Set – Volume 2)

GRUNWALD Armin
The Hermeneutic Side of Responsible Research and Innovation
(Responsible Research and Innovation Set – Volume 5)

LENOIR Virgil Cristian
Ethical Efficiency: Responsibility and Contingency
(Responsible Research and Innovation Set – Volume 1)

PELLÉ Sophie, REBER Bernard
From Ethical Review to Responsible Research and Innovation
(Responsible Research and Innovation Set – Volume 3)

REBER Bernard
Precautionary Principle, Pluralism and Deliberation: Science and Ethics
(Responsible Research and Innovation Set – Volume 4)

Printed and bound by CPI Group (UK) Ltd, Croydon, CR0 4YY
16/09/2026
14963078-0003